Relative Permeability of Petroleum Reservoirs

Authors

Mehdi Honarpour
Associate Professor of Petroleum Engineering
Department of Petroleum Engineering
Montana College of Mineral Science and Technology
Butte, Montana

Leonard Koederitz
Professor of Petroleum Engineering
Department of Petroleum Engineering
University of Missouri
Rolla, Missouri

A. Herbert Harvey
Chairman
Department of Petroleum Engineering
University of Missouri
Rolla, Missouri

CRC Press
Taylor & Francis Group
Boca Raton London New York

CRC Press is an imprint of the
Taylor & Francis Group, an **informa** business

First published 1986 by CRC Press
Taylor & Francis Group
6000 Broken Sound Parkway NW, Suite 300
Boca Raton, FL 33487-2742

Reissued 2018 by CRC Press

Library of Congress Cataloging in Publication Data
Main entry under title:

Relative permeability of petroleum reservoirs.

Bibliography: p.
Includes index.
1. Rocks—Permeability. 2. Oil reservoir engineering.
3. Carbonate reservoirs. I. Honarpour. Mehdi.
II. Koederitz, Leonard. III. Harvey, A. Herbert.
TN870.5.R45 1986 552'.5 85-7827
ISBN 0-8493-5739-X

A Library of Congress record exists under LC control number: 85007827

Publisher's Note
The publisher has gone to great lengths to ensure the quality of this reprint but points out that some imperfections in the original copies may be apparent.

Disclaimer
The publisher has made every effort to trace copyright holders and welcomes correspondence from those they have been unable to contact.

ISBN 13: 978-1-315-89722-6 (hbk)
ISBN 13: 978-1-351-07632-6 (ebk)

Visit the Taylor & Francis Web site at http://www.taylorandfrancis.com and the
CRC Press Web site at http://www.crcpress.com

PREFACE

In 1856 Henry P. Darcy determined that the rate of flow of water through a sand filter could be described by the equation

$$q = KA \frac{h_1 - h_2}{L}$$

where q represents the rate at which water flows downward through a vertical sand pack with cross-sectional area A and length L; the terms h_1 and h_2 represent hydrostatic heads at the inlet and outlet, respectively, of the sand filter, and K is a constant. Darcy's experiments were confined to the flow of water through sand packs which were 100% saturated with water.

Later investigators determined that Darcy's law could be modified to describe the flow of fluids other than water, and that the proportionality constant K could be replaced by k/μ, where k is a property of the porous material (permeability) and μ is a property of the fluid (viscosity). With this modification, Darcy's law may be written in a more general form as

$$v_s = \frac{k}{\mu}\left[\rho g \frac{dZ}{dS} - \frac{dP}{dS}\right]$$

where

S	=	Distance in direction of flow, which is taken as positive
v_s	=	Volume of flux across a unit area of the porous medium in unit time along flow path S
Z	=	Vertical coordinate, which is taken as positive downward
ρ	=	Density of the fluid
g	=	Gravitational acceleration
$\frac{dP}{dS}$	=	Pressure gradient along S at the point to which v_s refers

The volumetric flux v_s may be further defined as q/A, where q is the volumetric flow rate and A is the average cross-sectional area perpendicular to the lines of flow.

It can be shown that the permeability term which appears in Darcy's law has units of length squared. A porous material has a permeability of 1 D when a single-phase fluid with a viscosity of 1 cP completely saturates the pore space of the medium and will flow through it under viscous flow at the rate of 1 $cm^3/sec/cm^2$ cross-sectional area under a pressure gradient of 1 atm/cm. It is important to note the requirement that the flowing fluid must completely saturate the porous medium. Since this condition is seldom met in a hydrocarbon reservoir, it is evident that further modification of Darcy's law is needed if the law is to be applied to the flow of fluids in an oil or gas reservoir.

A more useful form of Darcy's law can be obtained if we assume that a rock which contains more than one fluid has an effective permeability to each fluid phase and that the effective permeability to each fluid is a function of its percentage saturation. The effective permeability of a rock to a fluid with which it is 100% saturated is equal to the absolute permeability of the rock. Effective permeability to each fluid phase is considered to be independent of the other fluid phases and the phases are considered to be immiscible.

If we define relative permeability as the ratio of effective permeability to absolute permeability, Darcy's law may be restated for a system which contains three fluid phases as follows:

$$v_{os} = \frac{k\,k_{ro}}{\mu_o}\left(\rho_o g \frac{dZ}{dS} - \frac{dP_o}{dS}\right)$$

$$v_{gs} = \frac{k\,k_{rg}}{\mu g}\left(\rho_g g \frac{dZ}{dS} - \frac{dP_g}{dS}\right)$$

$$v_{ws} = \frac{k\,k_{rw}}{\mu w}\left(\rho_w g \frac{dZ}{dS} - \frac{dP_w}{dS}\right)$$

where the subscripts o, g, and w represent oil, gas, and water, respectively. Note that k_{ro}, k_{rg}, and k_{rw} are the relative permeabilities to the three fluid phases at the respective saturations of the phases within the rock.

Darcy's law is the basis for almost all calculations of fluid flow within a hydrocarbon reservoir. In order to use the law, it is necessary to determine the relative permeability of the reservoir rock to each of the fluid phases; this determination must be made throughout the range of fluid saturations that will be encountered. The problems involved in measuring and predicting relative permeability have been studied by many investigators. A summary of the major results of this research is presented in the following chapters.

THE AUTHORS

Dr. Mehdi "Matt" Honarpour is an associate professor of petroleum engineering at the Montana College of Mineral Science and Technology, Butte, Montana. Dr. Honarpour obtained his B.S., M.S., and Ph.D. in petroleum engineering from the University of Missouri-Rolla. He has authored many publications in the area of reservoir engineering and core analysis. Dr. Honarpour has worked as reservoir engineer, research engineer, consultant, and teacher for the past 15 years. He is a member of several professional organizations, including the Society of Petroleum Engineers of AIME, the honorary society of Sigma Xi, Pi Epsilon Tau and Phi Kappa Phi.

Leonard F. Koederitz is a Professor of Petroleum Engineering at the University of Missouri-Rolla. He received B.S., M.S., and Ph.D. degrees from the University of Missouri-Rolla. Dr. Koederitz has worked for Atlantic-Richfield and previously served as Department Chairman at Rolla. He has authored or co-authored several technical publications and two texts related to reservoir engineering.

A. Herbert Harvey received B.S. and M.S. degrees from Colorado School of Mines and a Ph.D. degree from the University of Oklahoma. He has authored or co-authored numerous technical publications on topics related to the production of petroleum. Dr. Harvey is Chairman of both the Missouri Oil and Gas Council and the Petroleum Engineering Department at the University of Missouri-Rolla.

ACKNOWLEDGMENT

The authors wish to acknowledge the Society of Petroleum Engineers and the American Petroleum Institute for granting permission to use their publications. Special thanks are due J. Joseph of Flopetrol Johnston and A. Manjnath of Reservoir Inc. for their contributions and reviews throughout the writing of this book.

TABLE OF CONTENTS

Chapter 1

MEASUREMENT OF ROCK RELATIVE PERMEABILITY

I. INTRODUCTION

The relative permeability of a rock to each fluid phase can be measured in a core sample by either "steady-state" or "unsteady-state" methods. In the steady-state method, a fixed ratio of fluids is forced through the test sample until saturation and pressure equilibria are established. Numerous techniques have been successfully employed to obtain a uniform saturation. The primary concern in designing the experiment is to eliminate or reduce the saturation gradient which is caused by capillary pressure effects at the outflow boundary of the core. Steady-state methods are preferred to unsteady-state methods by some investigators for rocks of intermediate wettability,[1] although some difficulty has been reported in applying the Hassler steady-state method to this type of rock.[2]

In the capillary pressure method, only the nonwetting phase is injected into the core during the test. This fluid displaces the wetting phase and the saturations of both fluids change throughout the test. Unsteady-state techniques are now employed for most laboratory measurements of relative permeability.[3] Some of the more commonly used laboratory methods for measuring relative permeability are described below.

II. STEADY-STATE METHODS

A. Penn-State Method

This steady-state method for measuring relative permeability was designed by Morse et al.[4] and later modified by Osoba et al.,[5] Henderson and Yuster,[6] Caudle et al.,[7] and Geffen et al.[8] The version of the apparatus which was described by Geffen et al., is illustrated by Figure 1. In order to reduce end effects due to capillary forces, the sample to be tested is mounted between two rock samples which are similar to the test sample. This arrangement also promotes thorough mixing of the two fluid phases before they enter the test sample. The laboratory procedure is begun by saturating the sample with one fluid phase (such as water) and adjusting the flow rate of this phase through the sample until a predetermined pressure gradient is obtained. Injection of a second phase (such as a gas) is then begun at a low rate and flow of the first phase is reduced slightly so that the pressure differential across the system remains constant. After an equilibrium condition is reached, the two flow rates are recorded and the percentage saturation of each phase within the test sample is determined by removing the test sample from the assembly and weighing it. This procedure introduces a possible source of experimental error, since a small amount of fluid may be lost because of gas expansion and evaporation. One authority recommends that the core be weighed under oil, eliminating the problem of obtaining the same amount of liquid film on the surface of the core for each weighing.[3]

The estimation of water saturation by measuring electric resistivity is a faster procedure than weighing the core. However, the accuracy of saturations obtained by a resistivity measurement is questionable, since resistivity can be influenced by fluid distribution as well as fluid saturations. The four-electrode assembly which is illustrated by Figure 1 was used to investigate water saturation distribution and to determine when flow equilibrium has been attained. Other methods which have been used for *in situ* determination of fluid saturation in cores include measurement of electric capacitance, nuclear magnetic resonance, neutron scattering, X-ray absorption, gamma-ray absorption, volumetric balance, vacuum distillation, and microwave techniques.

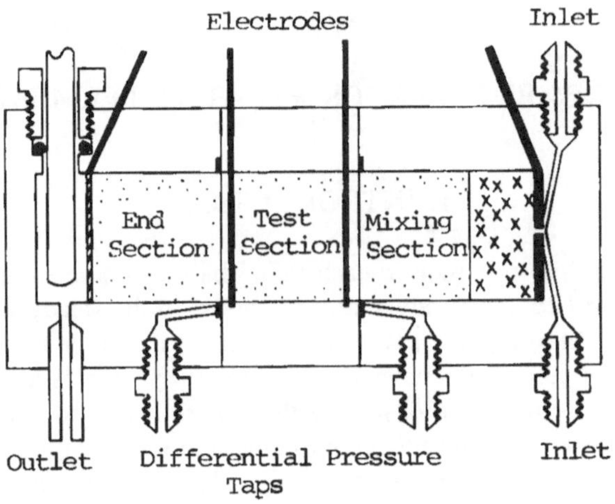

FIGURE 1. Three-section core assembly.[8]

After fluid saturation in the core has been determined, the Penn-State apparatus is reassembled, a new equilibrium condition is established at a higher flow rate for the second phase, and fluid saturations are determined as previously described. This procedure is repeated sequentially at higher saturations of the second phase until the complete relative permeability curve has been established.

The Penn-State method can be used to measure relative permeability at either increasing or decreasing saturations of the wetting phase and it can be applied to both liquid-liquid and gas-liquid systems. The direction of saturation change used in the laboratory should correspond to field conditions. Good capillary contact between the test sample and the adjacent downstream core is essential for accurate measurements and temperature must be held constant during the test. The time required for a test to reach an equilibrium condition may be 1 day or more.[3]

B. Single-Sample Dynamic Method

This technique for steady-state measurement of relative permeability was developed by Richardson et al.,[9] Josendal et al.,[10] and Loomis and Crowell.[11] The apparatus and experimental procedure differ from those used with the Penn-State technique primarily in the handling of end effects. Rather than using a test sample mounted between two core samples (as illustrated by Figure 1), the two fluid phases are injected simultaneously through a single core. End effects are minimized by using relatively high flow rates, so the region of high wetting-phase saturation at the outlet face of the core is small. The theory which was presented by Richardson et al. for describing the saturation distribution within the core may be developed as follows. From Darcy's law, the flow of two phases through a horizontal linear system can be described by the equations

$$-dP_{wt} = \frac{Q_{wt}\,\mu_{wt}\,dL}{k_{wt}\,A} \tag{1}$$

and

$$-dP_{n} = \frac{Q_{n}\,\mu_{n}\,dL}{k_{n}\,A} \tag{2}$$

where the subscripts wt and n denote the wetting and nonwetting phases, respectively. From the definition of capillary pressure, P_c, it follows that

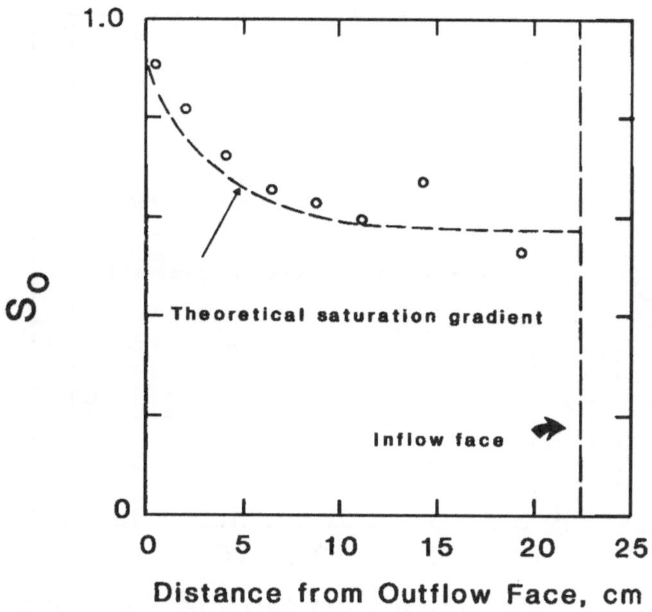

FIGURE 2. Comparison of saturation gradients at low flow rate.[9]

$$dP_c = dP_n - dP_{wt} \tag{3}$$

These three equations may be combined to obtain

$$\frac{dP_c}{dL} = \left(\frac{Q_{wt}\,\mu_{wt}}{k_{wt}} - \frac{Q_n\mu_n}{k_n}\right) \Big/ A \tag{4}$$

where dP_c/dL is the capillary pressure gradient within the core. Since

$$\frac{dP_c}{dL} = \frac{dP_c}{dS_{wt}}\,\frac{dS_{wt}}{dL} \tag{5}$$

it is evident that

$$\frac{dS_{wt}}{dL} = \frac{1}{A}\left(\frac{Q_{wt}\,\mu_{wt}}{k_{wt}} - \frac{Q_n\mu_n}{k_n}\right)\frac{1}{dP_c/dS_{wt}} \tag{6}$$

Richardson et al. concluded from experimental evidence that the nonwetting phase saturation at the discharge end of the core was at the equilibrium value, (i.e., the saturation at which the phase becomes mobile). With this boundary condition, Equation 6 can be integrated graphically to yield the distribution of wetting phase saturation throughout the core. If the flow rate is sufficiently high, the calculation indicates that this saturation is virtually constant from the inlet face to a region a few centimeters from the outlet. Within this region the wetting phase saturation increases to the equilibrium value at the outlet face. Both calculations and experimental evidence show that the region of high wetting-phase saturation at the discharge end of the core is larger at low flow rates than at high rates. Figure 2 illustrates the saturation distribution for a low flow rate and Figure 3 shows the distribution at a higher rate.

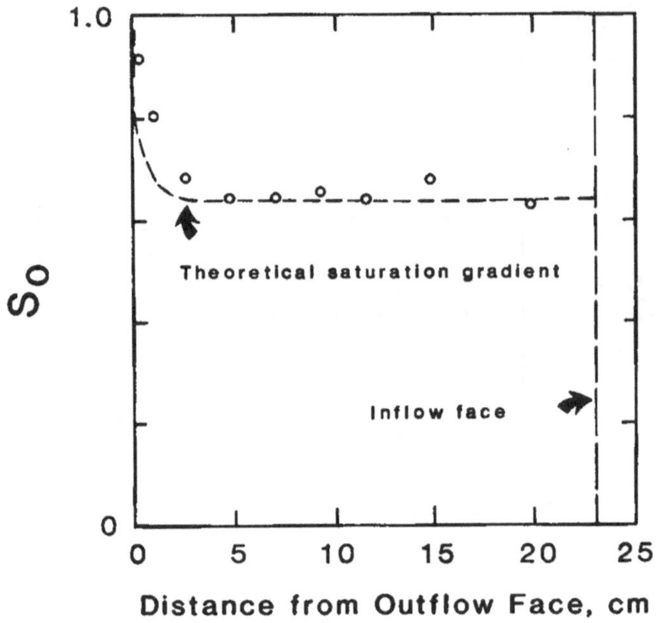

FIGURE 3. Comparison of saturation gradients at high flow rate.[9]

Although the flow rate must be high enough to control capillary pressure effects at the discharge end of the core, excessive rates must be avoided. Problems which can occur at very high rates include nonlaminar flow.

C. Stationary Fluid Methods

Leas et al.[12] described a technique for measuring permeability to gas with the liquid phase held stationary within the core by capillary forces. Very low gas flow rates must be used, so the liquid is not displaced during the test. This technique was modified slightly by Osoba et al.,[5] who held the liquid phase stationary within the core by means of barriers which were permeable to gas but not to the liquid. Rapoport and Leas[13] employed a similar technique using semipermeable barriers which held the gas phase stationary while allowing the liquid phase to flow. Corey et al.[14] extended the stationary fluid method to a three-phase system by using barriers which were permeable to water but impermeable to oil and gas. Osoba et al. observed that relative permeability to gas determined by the stationary liquid method was in good agreement with values measured by other techniques for some of the cases which were examined. However, they found that relative permeability to gas determined by the stationary liquid technique was generally lower than by other methods in the region of equilibrium gas saturation. This situation resulted in an equilibrium gas saturation value which was higher than obtained by the other methods used (Penn-State, Single-Sample Dynamic, and Hassler). Saraf and McCaffery consider the stationary fluid methods to be unrealistic, since all mobile fluids are not permitted to flow simultaneously during the test.[2]

D. Hassler Method

This is a steady-state method for relative permeability measurement which was described by Hassler[15] in 1944. The technique was later studied and modified by Gates and Lietz,[16] Brownscombe et al.,[17] Osoba et al.,[5] and Josendal et al.[10] The laboratory apparatus is illustrated by Figure 4. Semipermeable membranes are installed at each end of the Hassler test assembly. These membranes keep the two fluid phases separated at the inlet and outlet of the core, but allow both phases to flow simultaneously through the core. The pressure

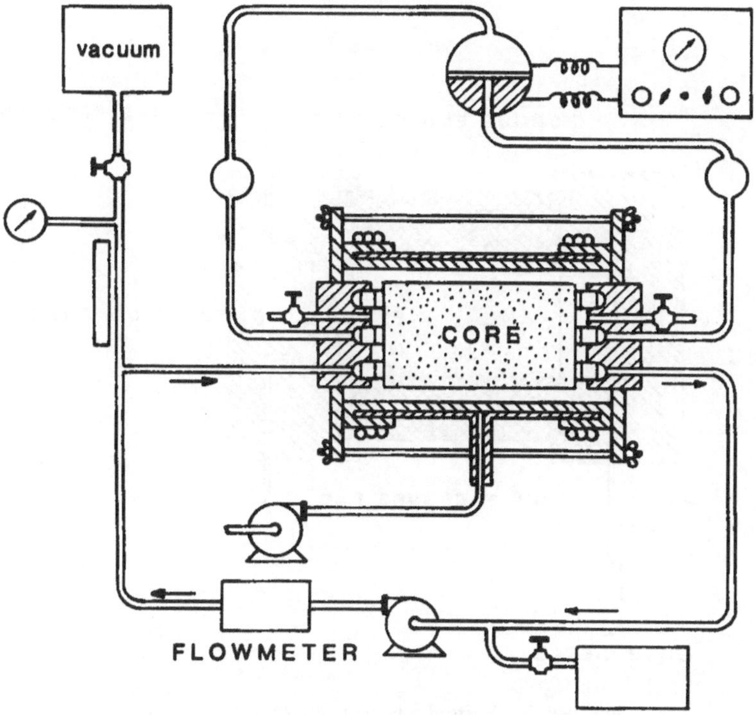

FIGURE 4. Two-phase relative permeability apparatus.[15]

in each fluid phase is measured separately through a semipermeable barrier. By adjusting the flow rate of the nonwetting phase, the pressure gradients in the two phases can be made equal, equalizing the capillary pressures at the inlet and outlet of the core. This procedure is designed to provide a uniform saturation throughout the length of the core, even at low flow rates, and thus eliminate the capillary end effect. The technique works well under conditions where the porous medium is strongly wet by one of the fluids, but some difficulty has been reported in using the procedure under conditions of intermediate wettability.[2,18] The Hassler method is not widely used at this time, since the data can be obtained more rapidly with other laboratory techniques.

E. Hafford Method

This steady-state technique was described by Richardson et al.[9] In this method the nonwetting phase is injected directly into the sample and the wetting phase is injected through a disc which is impermeable to the nonwetting phase. The central portion of the semipermeable disc is isolated from the remainder of the disc by a small metal sleeve, as illustrated by Figure 5. The central portion of the disc is used to measure the pressure in the wetting fluid at the inlet of the sample. The nonwetting fluid is injected directly into the sample and its pressure is measured through a standard pressure tap machined into the Lucite® surrounding the sample. The pressure difference between the wetting and the nonwetting fluid is a measure of the capillary pressure in the sample at the inflow end. The design of the Hafford apparatus facilitates investigation of boundary effects at the influx end of the core. The outflow boundary effect is minimized by using a high flow rate.

F. Dispersed Feed Method

This is a steady-state method for measuring relative permeability which was designed by Richardson et al.[9] The technique is similar to the Hafford and single-sample dynamic meth-

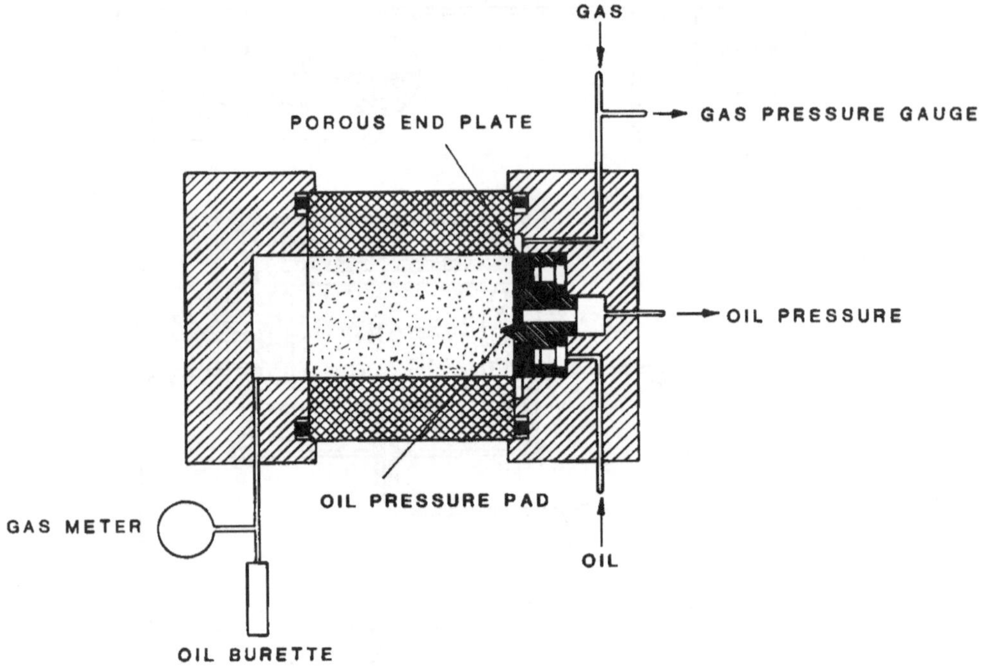

FIGURE 5. Hafford relative permeability apparatus.[9]

ods. In the dispersed feed method, the wetting fluid enters the test sample by first passing through a dispersing section, which is made of a porous material similar to the test sample. This material does not contain a device for measuring the input pressure of the wetting phase as does the Hafford apparatus. The dispersing section distributes the wetting fluid so that it enters the test sample more or less uniformly over the inlet face. The nonwetting phase is introduced into radial grooves which are machined into the outlet face of the dispersing section, at the junction between the dispersing material and the test sample. Pressure gradients used for the tests are high enough so the boundary effect at the outlet face of the core is not significant.

III. UNSTEADY-STATE METHODS

Unsteady-state relative permeability measurements can be made more rapidly than steady-state measurements, but the mathematical analysis of the unsteady-state procedure is more difficult. The theory developed by Buckley and Leverett[19] and extended by Welge[20] is generally used for the measurement of relative permeability under unsteady-state conditions. The mathematical basis for interpretation of the test data may be summarized as follows: Leverett[21] combined Darcy's law with a definition of capillary pressure in differential form to obtain

$$f_{w2} = \frac{1 + \dfrac{k_o}{q_t \mu_o}\left(\dfrac{\partial P_c}{\partial x} - g\,\Delta\rho\,\sin\theta\right)}{1 + \dfrac{k_o}{k_w} \cdot \dfrac{\mu_w}{\mu_o}} \tag{7}$$

where f_{w2} is the fraction water in the outlet stream; q_t is the superficial velocity of total fluid leaving the core; θ is the angle between direction x and horizontal; and $\Delta\rho$ is the density difference between displacing and displaced fluids. For the case of horizontal flow and negligible capillary pressure, Welge[20] showed that Equation 7 implies

$$S_{w,av} - S_{w2} = f_{o2}\, Q_w \tag{8}$$

where the subscript 2 denotes the outlet end of the core; $S_{w,av}$ is the average water saturation; and Q_w is the cumulative water injected, measured in pore volumes. Since Q_w and $S_{w,av}$ can be measured experimentally, f_{o2} (fraction oil in the outlet stream) can be determined from the slope of a plot of Q_w as a function of $S_{w,av}$. By definition

$$f_{o2} = q_o/(q_o + q_w) \tag{9}$$

By combining this equation with Darcy's law, it can be shown that

$$f_{o2} = \frac{1}{1 + \dfrac{\mu_o/k_{ro}}{\mu_w/k_{rw}}} \tag{10}$$

Since μ_o and μ_w are known, the relative permeability ratio k_{ro}/k_{rw} can be determined from Equation 10. A similar expression can be derived for the case of gas displacing oil.

The work of Welge was extended by Johnson et al.[22] to obtain a technique (sometimes called the JBN method) for calculating individual phase relative permeabilities from unsteady-state test data. The equations which were derived are

$$k_{ro} = \frac{f_{o2}}{d\left(\dfrac{1}{Q_w I_r}\right) \Big/ d\left(\dfrac{1}{Q_w}\right)} \tag{11}$$

and

$$k_{rw} = \frac{f_{w2}}{f_{o2}} \frac{\mu_w}{\mu_o} k_{ro} \tag{12}$$

where I_r, the relative injectivity, is defined as

$$I_r = \frac{\text{injectivity}}{\text{initial injectivity}} = \frac{(q_{wi}/\Delta p)}{(q_{wi}/\Delta p) \text{ at start of injection}} \tag{13}$$

A graphical technique for solving Equations 11 and 12 is illustrated in Reference 23. Relationships describing relative permeabilities in a gas-oil system may be obtained by replacing the subscript "w" with "g" in Equations 11, 12, and 13.

In designing experiments to determine relative permeability by the unsteady-state method, it is necessary that:

1. The pressure gradient be large enough to minimize capillary pressure effects.
2. The pressure differential across the core be sufficiently small compared with total operating pressure so that compressibility effects are insignificant.
3. The core be homogeneous.
4. The driving force and fluid properties be held constant during the test.[2]

Laboratory equipment is available for making the unsteady-state measurements under simulated reservoir conditions.[24]

In addition to the JBN method, several alternative techniques for determining relative permeability from unsteady-state test data have been proposed. Saraf and McCaffery[2] developed a procedure for obtaining relative permeability curves from two parameters determined by least squares fit of oil recovery and pressure data. The technique is believed to be superior to the JBN method for heterogeneous carbonate cores. Jones and Roszelle[25] developed a graphical technique for evaluation of individual phase relative permeabilities from displacement experimental data which are linearly scalable. Chavent et al. described a method for determining two-phase relative permeability and capillary pressure from two sets of displacement experiments, one set conducted at a high flow rate and the other at a rate representative of reservoir conditions. The theory of Welge was extended by Sarem to describe relative permeabilities in a system containing three fluid phases. Sarem employed a simplifying assumption that the relative permeability to each phase depends only on its own saturation, and the validity of this assumption (particularly with respect to the oil phase) has been questioned.[2]

Unsteady-state relative permeability measurements are frequently used to determine the ratios k_w/k_o, k_g/k_o, and k_g/k_w. The ratio k_w/k_o is used to predict the performance of reservoirs which are produced by waterflood or natural water drive; k_g/k_o is employed to estimate the production which will be obtained from recovery processes where oil is displaced by gas, such as gas injection or solution gas drive. An important use of the ratio k_g/k_w is in the prediction of performance of natural gas storage wells, where gas is injected into an aquifier. The ratios k_w/k_o, k_g/k_o, and k_g/k_w are usually measured in a system which contains only the two fluids for which the relative permeability ratio is to be determined. It is believed that the connate water in the reservoir may have an influence on k_g/k_o, expecially in sandstones which contain hydratable clay minerals and in low permeability rock. For these types of reservoirs it may be advisable to measure k_g/k_o in cores which contain an immobile water saturation.[24]

IV. CAPILLARY PRESSURE METHODS

The techniques which are used for calculating relative permeability from capillary pressure data were developed for drainage situations, where a nonwetting phase (gas) displaces a wetting phase (oil or water). Therefore use of the techniques is generally limited to gas-oil or gas-water systems, where the reservoir is produced by a drainage process. Although it is possible to calculate relative permeabilities in a water-oil system from capillary pressure data, accuracy of this technique is uncertain; the displacement of oil by water in a water-wet rock is an imbibition process rather than a drainage process.

Although capillary pressure techniques are not usually the preferred methods for generating relative permeability data, the methods are useful for obtaining gas-oil or gas-water relative permeabilities when rock samples are too small for flow tests but large enough for mercury injection. The techniques are also useful in rock which has such low permeability that flow tests are impractical and for instances where capillary pressure data have been measured but a sample of the rock is not available for measuring relative permeability. Another use which has been suggested for the capillary pressure techniques is in estimating k_g/k_o ratios for retrograde gas condensate reservoirs, where oil saturation increases as pressure decreases, with an initial oil saturation which may be as low as zero. The capillary pressure methods are recommended for this situation because the conventional unsteady-state test is not designed for very low oil saturations.

Data obtained by mercury injection are customarily used when relative permeability is estimated by the capillary pressure technique. The core is evacuated and mercury (which is the nonwetting phase) is injected in measured increments at increasing pressures. Approximately 20 data points are obtained in a typical laboratory test designed to yield the complete capillary pressure curve, which is required for calculating relative permeability by the methods described below.

Several investigators have developed equations for estimating relative permeability from capillary pressure data. Purcell[29] presented the equations

$$k_{rwt} = \frac{\int_0^{S_{wi}} dS/p_c^n}{\int_0^1 dS/p_c^n} \tag{14}$$

and

$$k_{rnwt} = \frac{\int_{S_{wi}}^1 dS/p_c^n}{\int_0^1 dS/p_c^n} \tag{15}$$

where the subscripts wt and nwt denote the wetting and nonwetting phases, respectively, and n has a value of 2.0. Fatt and Dykstra[30] developed similar equations with n equal to 3.0.

A slightly different result is obtained by combining the equations developed by Burdine[31] with the work of Purcell.[29] The results are

$$k_{rwt} = \left(\frac{S_L - S_{wi}}{1 - S_{wi}}\right)^2 \frac{\int_{S_{wi}}^{S_L} dS_w/P_c^2}{\int_{S_{wi}}^1 dS_w/P_c^2} \tag{16}$$

$$k_{rnwt} = \left(\frac{1 - S_L}{1 - S_{wi}}\right)^2 \frac{\int_{S_L}^1 dS_w/P_c^2}{\int_{S_{wi}}^1 dS_w/P_c^2} \tag{17}$$

where S_L is the total liquid saturation.

V. CENTRIFUGE METHODS

Centrifuge techniques for measuring relative permeability involve monitoring liquids produced from rock samples which were initially saturated uniformly with one or two phases. Liquids are collected in transparent tubes connected to the rock sample holders and production is monitored throughout the test. Mathematical techniques for deriving relative permeability data from these measurements are described in References 26, 27, and 28.

Although the centrifuge methods have not been widely used, they do offer some advantages over alternative techniques. The centrifuge methods are substantially faster than the steady-state techniques and they apparently are not subject to the viscous fingering problems which sometimes interfere with the unsteady-state measurements. On the other hand, the centrifuge methods are subject to capillary end effect problems and they do not provide a means for determining relative permeability to the invading phase.

O'Mera and Lease[28] describe an automated centrifuge which employs a photodiode array in conjunction with a microcomputer to image and identify liquids produced during the test.

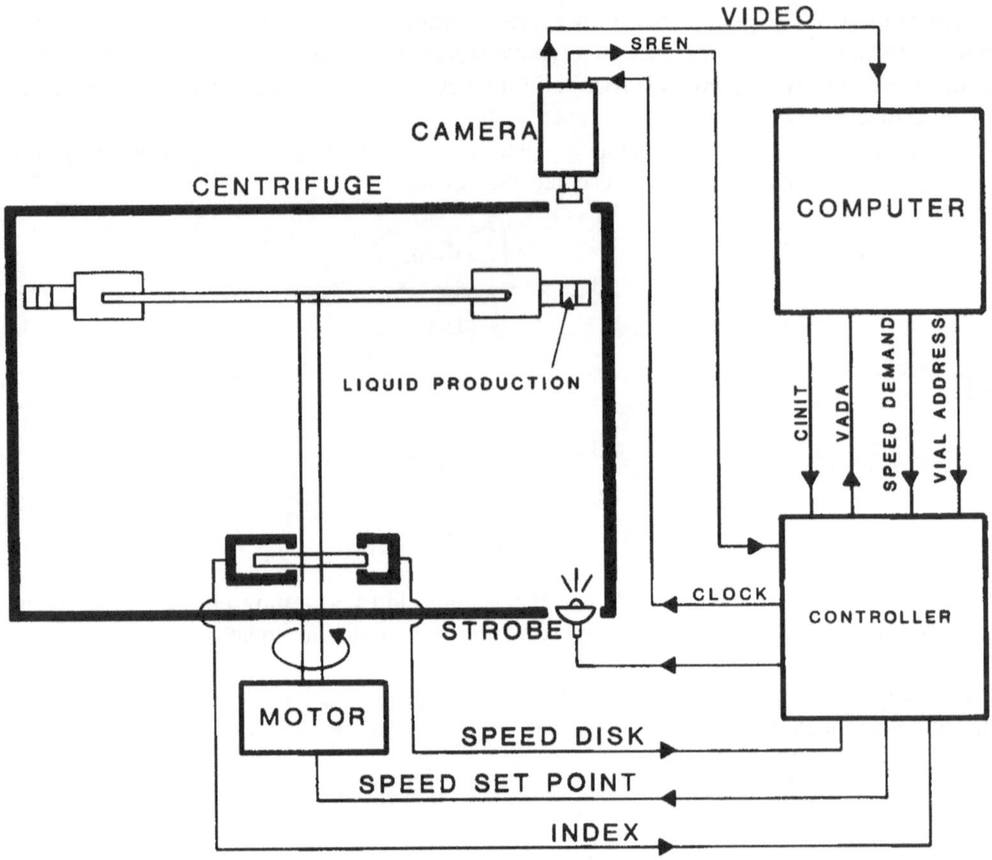

FIGURE 6. Automated centrifuge system.[28]

Stroboscopic lights are located below the rotating tubes and movement of fluid interfaces is monitored by the transmitted light. Fluid collection tubes are square in cross section, since a cylindrical tube would act as a lens and concentrate the light in a narrow band along the major axis of the tube. A schematic diagram of the apparatus is shown by Figure 6.

VI. CALCULATION FROM FIELD DATA

It is possible to calculate relative permeability ratios directly from field data.[23] In making the computation it is necessary to recognize that part of the gas which is produced at the surface was dissolved within the liquid phase in the reservoir. Thus;

$$(\text{produced gas}) = (\text{free gas}) + (\text{solution gas}) \tag{18}$$

If we consider the flow of free gas in the reservoir, Darcy's law for a radial system may be written

$$q_{g,\text{free}} = 7.08 \, \frac{k_g h}{\mu_g B_g} \, \frac{P_e - P_w}{\ln (r_e/r_w)} \tag{19}$$

1	2	3	4	5	6	7	8	9
P psia	Np STB	Rp SCF/STB	$\frac{N-Np}{Np}$	$\frac{B_o}{B_{ob}}$	S_L = Sw + (1-Sw).(4)(5)	$\frac{Bg\,\mu g}{5.615\,B_o\mu_o}$	Rs SCF/STB	K_g/K_o = (3-8)(7)

FIGURE 7. Calculation of gas-oil relative permeability values from production data.

Similarly, the rate of oil flow in the same system is

$$q_o = 7.08 \frac{k_o h}{\mu_o B_o} \frac{P_e - P_w}{\ln (r_e/r_w)} \tag{20}$$

where r_w is the well radius and r_e is the radius of the external boundary of the area drained by the well. B_o and B_g are the oil and gas formation volume factors, respectively. The ratio of free gas to oil is obtained by dividing Equation 19 by Equation 20. If we express R_p, cumulative gas/oil ratio and R_s, solution gas/oil ratio, in terms of standard cubic foot per stock tank barrel, Equation 18 implies

$$R_p = 5.615 \frac{k_g}{k_o} \frac{\mu_o}{\mu_g} \frac{B_o}{B_g} + R_s \tag{21}$$

Thus, the relative permeability ratio is given by

$$\frac{k_g}{k_o} = \frac{(R_p - R_s)}{5.615} \frac{B_g}{B_o} \frac{\mu_g}{\mu_o} \tag{22}$$

The oil saturation which corresponds to this relative permeability ratio may be determined from a material balance. If we assume there is no water influx, no water production, no fluid injection, and no gas cap, the material balance equation may be written

$$S_o = \left(1 - \frac{N}{N_p}\right) \frac{B_o}{B_{oi}} (1 - S_w) \tag{23}$$

where minor effects such as change in reservoir pore volume have been assumed negligible. In Equation 23 the symbol N denotes initial stock tank barrels of oil in place; N_p is number of stock tank barrels of oil produced; and B_{oi} is the ratio of the oil volume at initial reservoir conditions to oil volume at standard conditions.

If total liquid saturation in the reservoir is expressed as

$$S_L = S_w + (1 - S_w) \left(\frac{N - N_p}{N_p}\right) \left(\frac{B_o}{B_{oi}}\right) \tag{24}$$

then the relative permeability curve may be obtained by plotting k_g/k_o from Equation 22 as a function of S_L from Equation 24. Figure 7 illustrates a convenient format for tabulating the data. The curve is prepared by plotting column 9 as a function of column 6 on semilog paper, with k_g/k_o on the logarithmic scale. The technique is useful even if only a few high-liquid-saturation data points can be plotted. These k_g/k_o values can be used to verify the accuracy of relative permeability predicted by empirical or laboratory techniques.

Poor agreement between relative permeability determined from production data and from laboratory experiments is not uncommon. The causes of these discrepancies may include the following:

1. The core on which relative permeability is measured may not be representative of the reservoir in regard to such factors as fluid distributions, secondary porosity, etc.
2. The technique customarily used to compute relative permeability from field data does not allow for the pressure and saturation gradients which are present in the reservoir, nor does it allow for the fact that wells may be producing from several strata which are at various stages of depletion.
3. The usual technique for calculating relative permeability from field data assumes that R_p at any pressure is constant throughout the oil zone. This assumption can lead to computational errors if gravitational effects within the reservoir are significant.

When relative permeability to water is computed from field data, a common source of error is the production of water from some source other than the hydrocarbon reservoir. These possible sources of extraneous water include casing leaks, fractures that extend from the hydrocarbon zone into an aquifer, etc.

REFERENCES

1. **Gorinik, B. and Roebuck, J. F.,** *Formation Evaluation through Extensive Use of Core Analysis,* Core Laboratories, Inc., Dallas, Tex., 1979.
2. **Saraf, D. N. and McCaffery, F. G.,** Two- and Three-Phase Relative Permeabilities: a Review, Petroleum Recovery Institute Report #81-8, Calgary, Alberta, Canada, 1982.
3. **Mungan, N.,** Petroleum Consultants Ltd., personal communication, 1982.
4. **Morse, R. A., Terwilliger, P. L., and Yuster, S. T.,** Relative permeability measurements on small samples, *Oil Gas J.,* 46, 109, 1947.
5. **Osoba, J. S., Richardson, J. G., Kerver, J. K., Hafford, J. A., and Blair, P. M.,** Laboratory relative permeability measurements, *Trans. AIME,* 192, 47, 1951.
6. **Henderson, J. H. and Yuster, S. T.,** Relative permeability study, *World Oil,* 3, 139, 1948.
7. **Caudle, B. H., Slobod, R. L., and Brownscombe, E. R. W.,** Further developments in the laboratory determination of relative permeability, *Trans. AIME,* 192, 145, 1951.
8. **Geffen, T. M., Owens, W. W., Parrish, D. R., and Morse, R. A.,** Experimental investigation of factors affecting laboratory relative permeability measurements, *Trans. AIME,* 192, 99, 1951.
9. **Richardson, J. G., Kerver, J. K., Hafford, J. A., and Osoba, J. S.,** Laboratory determination of relative permeability, *Trans. AIME,* 195, 187, 1952.
10. **Josendal, V. A., Sandiford, B. B., and Wilson, J. W.,** Improved multiphase flow studies employing radioactive tracers, *Trans. AIME,* 195, 65, 1952.
11. **Loomis, A. G. and Crowell, D. C.,** Relative Permeability Studies: Gas-Oil and Water-Oil Systems, U.S. Bureau of Mines Bulletin BarHeuillr, Okla., 1962, 599.
12. **Leas, W. J., Jenks, L. H., and Russell, Charles D.,** Relative permeability to gas, *Trans. AIME,* 189, 65, 1950.
13. **Rapoport, L. A. and Leas, W. J.,** Relative permeability to liquid in liquid-gas systems, *Trans. AIME,* 192, 83, 1951.
14. **Corey, A. T., Rathjens, C. H., Henderson, J. H., and Wyllie, M. R. J.,** Three-phase relative permeability, *J. Pet. Technol.,* Nov., 63, 1956.
15. **Hassler, G. L.,** U.S. Patent 2,345,935, 1944.
16. **Gates, J. I. and Leitz, W. T.,** Relative permeabilities of California cores by the capillary-pressure method, *Drilling and Production Practices,* American Petroleum Institute, Washington, D.C. 1950, 285.
17. **Brownscombe, E. R., Slobod, R. L., and Caudle, B. H.,** Laboratory determination of relative permeability, *Oil Gas J.,* 48, 98, 1950.
18. **Rose, W.,** Some problems in applying the Hassler relative permeability method, *J. Pet. Technol.,* 8, 1161, 1980.
19. **Buckley, S. E. and Leverett, M. C.,** Mechanism of fluid displacement in sands, *Trans. AIME,* 146, 107, 1942.
20. **Welge, H. J.,** A simplified method for computing recovery by gas or water drive, *Trans. AIME,* 195, 91, 1952.
21. **Leverett, M. C.,** Capillary behavior in porous solids, *Trans. AIME,* 142, 152, 1941.

22. **Johnson, E. F., Bossler, D. P., and Naumann, V. O.,** Calculation of relative permeability from displacement experiments, *Trans. AIME*, 216, 370, 1959.
23. **Crichlow, H. B., Ed.,** *Modern Reservoir Engineering — A Simulation Approach*, Prentice-Hall, Englewood Cliffs, 1977, chap. 7.
24. Special Core Analysis, Core Laboratories, Inc., Dallas, 1976.
25. **Jones, S. C. and Roszelle, W. O.,** Graphical techniques for determining relative permeability from displacement experiments, *J. Pet. Technol.*, 5, 807, 1978.
26. **Slobod, R. L., Chambers, A., and Prehn, W. L.,** Use of centrifuge for determining connate water, residual oil, and capillary pressure curves of small core samples, *Trans. AIME*, 192, 127, 1952.
27. **Van Spronsen, E.,** Three-phase relative permeability measurements using the Centrifuge Method, Paper SPE/DOE 10688 presented at the Third Joint Symposium, Tulsa, Okla., 1982.
28. **O'Mera, D. J., Jr. and Lease, W. O.,** Multiphase relative permeability measurements using an automated centrifuge, Paper SPE 12128 presented at the SPE 58th Annual Technical Conference and Exhibition, San Francisco, 1983.
29. **Purcell, W. R.,** Capillary pressures — their measurement using mercury and the calculation of permeability therefrom, *Trans. AIME*, 186, 39, 1949.
30. **Fatt, I. and Dyksta, H.,** Relative permeability studies, *Trans. AIME*, 192, 41, 1951.
31. **Burdine, N. T.,** Relative Permeability Calculations from Pore Size Distribution Data, *Trans. AIME*, 198, 71, 1953.

Chapter 2

TWO-PHASE RELATIVE PERMEABILITY

I. INTRODUCTION

Direct experimental measurement to determine relative permeability of porous rock has long been recorded in petroleum related literature. However, empirical methods for determining relative permeability are becoming more widely used, particularly with the advent of digital reservoir simulators. The general shape of the relative permeability curves may be approximated by the following equations: $k_{rw} = A(S_w)^n$; $k_{ro} = B(1 - S_w)^m$; where A, B, n, and m are constants.

Most relative permeability mathematical models may be classified under one of four categories:

Capillary models — Are based on the assumption that a porous medium consists of a bundle of capillary tubes of various diameters with a fluid path length longer than the sample. Capillary models ignore the interconnected nature of porous media and frequently do not provide realistic results.

Statistical models — Are also based on the modeling of porous media by a bundle of capillary tubes with various diameters distributed randomly. The models may be described as being divided into a large number of thin slices by planes perpendicular to the axes of the tubes. The slices are imagined to be rearranged and reassembled randomly. Again, statistical models have the same deficiency of not being able to model the interconnected nature of porous media.

Empirical models — Are based on proposed empirical relationships describing experimentally determined relative permeabilities and in general, have provided the most successful approximations.

Network models — Are frequently based on the modeling of fluid flow in porous media using a network of electric resistors as an analog computer. Network models are probably the best tools for understanding fluid flow in porous media.[1,44]

The hydrodynamic laws generally bear little use in the solution of problems concerning single-phase fluid flow through porous media, let alone multiphase fluid flow, due to the complexity of the porous system. One of the early attempts to relate several laboratory-measured parameters to rock permeability was the Kozeny-Carmen equation.[2] This equation expresses the permeability of a porous material as a function of the product of the effective path length of the flowing fluid and the mean hydraulic radius of the channels through which the fluid flows.

Purcell[3] formulated an equation for the permeability of a porous system in terms of the porosity and capillary pressure desaturation curve of that system by simply considering the porous medium as a bundle of capillary tubes of varying sizes.

Several authors[4-16] adapted the relations developed by Kozeny-Carmen and Purcell to the computation of relative permeability. They all proposed models on the basis of the assumption that a porous medium consists of a bundle of capillaries in order to apply Darcy's and Poiseuille's equations in their derivations. They used the tortuosity concept or texture parameters to take into account the tortuous path of the flow channels as opposed to the concept of capillary tubes. They tried to determine tortuosity empirically in order to obtain a close approximation of experimental data.

II. RAPOPORT AND LEAS

Rapoport and Leas[9] presented two equations for relative permeability to the wetting phase.

These equations were based on surface energy relationships and the Kozeny-Carmen equation. The equations were presented as defining limits for wetting-phase relative permeability.

The maximum and minimum wetting-phase relative permeability presented by Rapoport and Leas are

$$k_{rwt}(max) = \frac{\left(\frac{S_{wt} - S_m}{1 - S_m}\right)^3 \left(\int_1^{S_m} P_c\, dS\right)^2}{\int_{S_{wt}}^{S_m} P_c\, dS \dfrac{\int_1^{S_{wt}} P_c\, dS}{\left(\frac{1-\phi}{\phi S_{wt}}\right)\left(\frac{P_c(S_{wt} - S_m)}{\int_{S_{wt}}^{S_m} P_c\, dS}\right)^2}} \quad (1)$$

and

$$k_{rwt}(min) = \left(\frac{S_{wt} - S_m}{1 - S_m}\right)^3 \frac{\int_1^{S_{wt}} P_c\, dS}{\int_1^{S_m} P_c\, dS + \int_1^{S_{wt}} P_c\, dS} \quad (2)$$

where S_m represents the minimum irreducible saturation of the wetting phase from a drainage capillary pressure curve, expressed as a fraction; S_{wt} represents the saturation of the wetting phase for which the wetting-phase relative permeability is evaluated, expressed as a fraction; P_c represents the drainage capillary pressure expressed in psi and ϕ represents the porosity expressed as a fraction.

III. GATES, LIETZ, AND FULCHER

Gates and Lietz[8] developed the following expression based on Purcell's model for wetting-phase relative permeability:

$$k_{rwt} = \frac{\int_0^{S_{wt}} \frac{dS}{P_c^2}}{\int_0^1 \frac{dS}{P_c^2}} \quad (3)$$

Fulcher et al.,[45] have investigated the influence of capillary number (ratio of viscous to capillary forces) on two-phase oil-water relative permeability curves.

IV. FATT, DYKSTRA, AND BURDINE

Fatt and Dykstra[11] developed an expression for relative permeability following the basic method of Purcell for calculating the permeability of a porous medium. They considered a lithology factor (a correction for deviation of the path length from the length of the porous medium) to be a function of saturation. They assumed that the radius of the path of the conducting pores was related to the lithology factor, λ, by the equation:

$$\lambda = \frac{a}{r^b} \quad (4)$$

Table 1
CALCULATION OF WETTING-PHASE RELATIVE PERMEABILITY BASED ON THE FATT AND DYKSTRA EQUATION

S_w, %	P_c, cm Hg	$1/P_c^3$, (cm Hg)$^{-3}$	Area from 0 to S_w, in.^{2}	k_{rwt}, %
100	4.0	0.0156	11.25	100.0
90	4.5	0.0110	7.88	70.0[a]
80	5.0	0.0080	5.54	49.2[b]
70	5.5	0.0060	3.80	33.8
60	6.0	0.0046	2.49	22.1
50	6.7	0.0033	1.50	13.3
40	7.5	0.0024	0.75	6.7
30	8.7	0.0015	0.30	2.7
20	13.0	0.0005	0.20	0.4

[a] 7.88/11.25 × 100 = 70.0.
[b] 5.54/11.25 × 100 = 49.2.

where r represents the radius of a pore, a and b represent material constants, and λ is a function of saturation.

The equation for the wetting-phase relative permeability, k_{rwt}, reported by Fatt and Dykstra is

$$k_{rwt} = \frac{\int_0^{S_{wt}} \frac{dS}{P_c^{2(1+b)}}}{\int_0^1 \frac{dS}{P_c^{2(1+b)}}} \tag{5}$$

Fatt and Dykstra found good agreement with observed data when b = $^1/_2$, reducing Equation 5 to

$$k_{rwt} = \frac{\int_0^{S_{wt}} \frac{dS}{P_c^3}}{\int_0^1 \frac{dS}{P_c^3}} \tag{6}$$

They stated that their equation fit their own data as well as the data of Gates and Lietz more accurately than other proposed models.

The procedure for the calculation of relative permeability from capillary pressure data is illustrated by Table 1 and the results are shown in Figures 1 and 2.

Burdine[13] reported equations for computing relative permeability for both the wetting and nonwetting phases. His equations can be shown to reduce to a form similar to those developed by Purcell. Burdine's contribution is principally useful in handling tortuosity.

Defining the tortuosity factor for a pore as λ when the porous medium is saturated with only one fluid and using the symbol λ_{wt} for the wetting-phase tortuosity factor when two phases are present, a tortuosity ratio can be defined as

$$\lambda_{rwt} = \frac{\lambda}{\lambda_{wt}} \tag{7}$$

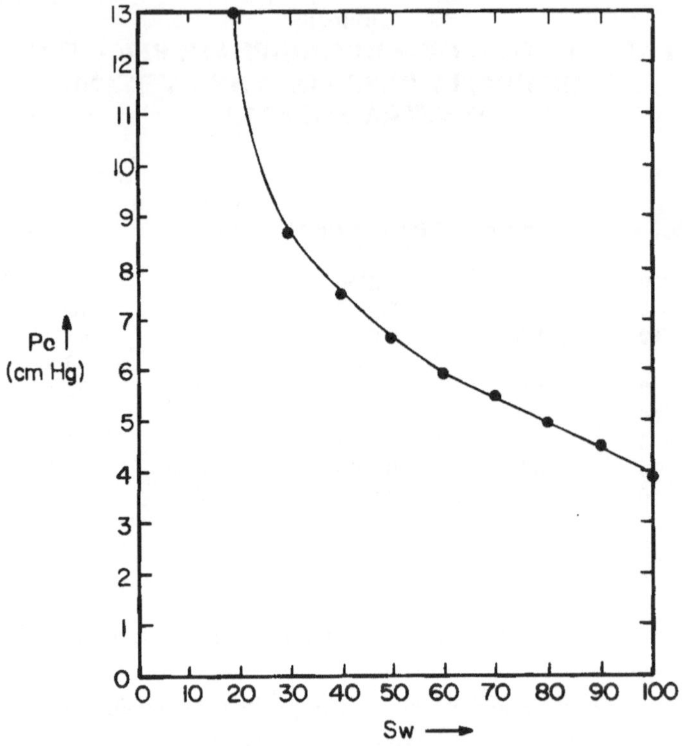

FIGURE 1. Capillary pressure as a function of water saturation.

then

$$k_{rwt} = \frac{\int_0^{S_{wt}} (\lambda_{rwt})^2 dS/(\lambda)^2(P_c)^2}{\int_0^1 dS/(\lambda)^2(P_c)^2} \quad (8)$$

If λ is a constant for the porous medium and λ_{rwt} depends only on the final saturation, then

$$k_{rwt} = (\lambda_{rwt})^2 \frac{\int_0^{S_{wt}} dS/(P_c)^2}{\int_0^1 dS/(P_c)^2} \quad (9)$$

In a similar fashion, the relative permeability to the nonwetting phase can be expressed utilizing a nonwetting-phase tortuosity ratio, λ_{rnwt},

$$k_{rnwt} = (\lambda_{rnwt})^2 \frac{\int_{S_{wt}}^1 dS/(P_c)^2}{\int_0^1 dS/(P_c)^2} \quad (10)$$

Burdine has shown that

$$\lambda_{rwt} = \frac{S_{wt} - S_m}{1 - S_m} \quad (11)$$

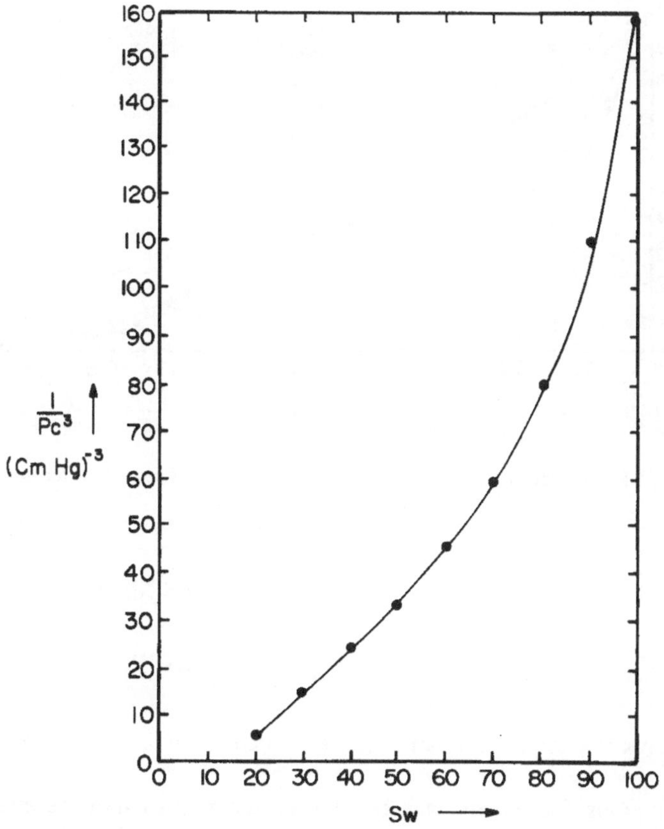

FIGURE 2. Reciprocal of (capillary pressure)^3 as a function of water saturation.

where S_m represents the minimum wetting-phase saturation from a capillary-pressure curve. The relative permeability is assumed to approach zero at this saturation. The nonwetting phase tortuosity can be approximated by

$$\lambda_{rnwt} = \frac{S_{nwt} - S_e}{1 - S_m - S_e} \tag{12}$$

where S_e is the equilibrium saturation to the nonwetting phase.

The expression for the wetting phase (Equation 9) fit the data presented much better than the expression for the nonwetting phase (Equation 10).

V. WYLLIE, SPRANGLER, AND GARDNER

Wyllie and Sprangler[12] reported equations similar to those presented by Burdine for computing oil and gas relative permeability. Their equations can be expressed as follows:

$$k_{ro} = \left(\frac{S_o}{1 - S_{wi}}\right)^2 \frac{\int_0^{S_o} dS_o/P_c^2}{\int_0^1 dS_o/P_c^2} \tag{13}$$

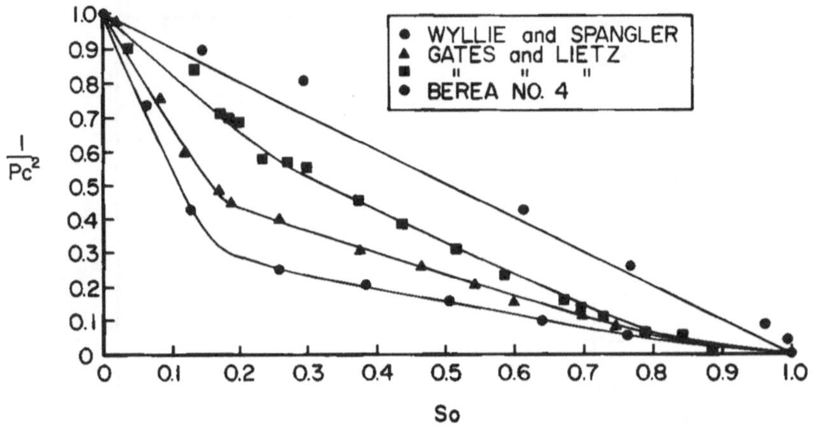

FIGURE 3. Reciprocal of (capillary pressure)^2 as a function of saturation for normalized data.[17]

$$k_{rg} = \left(1 - \frac{S_o}{S_m - S_{wi}}\right)^2 \frac{\int_{S_o}^{1} dS_o/P_c^2}{\int_{0}^{1} dS_o/P_c^2} \quad (14)$$

where S_m represents the lowest oil saturation at which the gas phase is discontinuous: $S_m = (1 - S_{gc})$.

The above equations for oil and gas relative permeabilities may be evaluated when a reliable drainage capillary pressure curve of the porous medium is available, so that a plot of $1/P_c^2$ as a function of oil saturation can be constructed. Obviously, reliable values of S_m and S_{or} are also needed for the oil and gas relative permeability evaluation. Figure 3 shows some examples of $1/P_c^2$ vs. saturation curves.[17]

Wyllie and Gardner[17] developed equations for oil and gas relative permeabilities in the presence of an irreducible water saturation, with the water considered as part of the rock matrix:

$$k_{ro} = \left(\frac{S_L - S_{wi}}{1 - S_{wi}}\right)^2 \frac{\int_{S_{wi}}^{S_L} \frac{dS_w}{P_c^2}}{\int_{S_{wi}}^{1} \frac{dS_w}{P_c^2}} \quad (15)$$

$$k_{rg} = \left(\frac{1 - S_L}{1 - S_{wi}}\right)^2 \frac{\int_{S_L}^{1} \frac{dS_w}{P_c^2}}{\int_{S_{wi}}^{1} \frac{dS_w}{P_c^2}} \quad (16)$$

where S_L represents total liquid saturation. Note that these equations may be applied only when the water saturation is at the irreducible level.

VI. TIMMERMAN, COREY, AND JOHNSON

Timmerman[18] suggests the following equations based on the water-oil drainage capillary pressure, for the calculation of low values of water-oil relative permeability.

Wetting-Phase Drainage Process:

$$k_{ro} = S_o \left[\frac{\int_0^{S_o} \frac{dS}{P_c^2}}{\int_0^1 \frac{dS}{P_c^2}} \right]^{2.5} \quad \begin{matrix} \text{Injection Curve} \\ \\ \text{Injection Curve} \end{matrix} \tag{17}$$

$$k_{rw} = S_w \left[\frac{\int_{S_o}^{1} \frac{dS}{P_c^2}}{\int_0^1 \frac{dS}{P_c^2}} \right]^{2.5} \quad \begin{matrix} \text{Injection Curve} \\ \\ \text{Injection Curve} \end{matrix} \tag{18}$$

Wetting-Phase Imbibition Process:

$$k_{ro} = S_o \left[\frac{\int_0^{S_o} \frac{dS}{P_c^2}}{\int_0^1 \frac{dS}{P_c^2}} \right]^{2.5} \quad \begin{matrix} \text{Injection Curve} \\ \\ \text{Injection Curve} \end{matrix} \tag{19}$$

$$k_{ro} = S_o \left[\frac{\int_{S_o}^{1} \frac{dS}{P_c^2}}{\int_0^1 \frac{dS}{P_c^2}} \right]^{2.5} \quad \begin{matrix} \text{Trap-Hysteresis Curve} \\ \\ \text{Injection Curve} \end{matrix} \tag{20}$$

Corey[19] combined the work of Purcell[3] and Burdine[13] into a form that has considerable utility and is widely accepted for its simplicity. It requires limited input data (since residual saturation is the only parameter needed to develop a set of relative permeability curves) and it is fairly accurate for consolidated porous media with intergranular porosity. Corey's equations are often used for calculation of relative permeability in reservoirs subject to a drainage process or external gas drive. His method of calculation was derived from capillary pressure concepts and the fact that for certain cases, $1/P_c^2$ is approximately a linear function of the effective saturation over a considerable range of saturations; i.e., $1/P_c^2 = C\,[(S_o - S_{or})/(1 - S_{or})]$ where C is a constant and S_o is an oil saturation greater than S_{or}. On the basis of this observation and the findings of Burdine[13] concerning the nature of the tortuosity-saturation function, the following expressions were derived:

$$k_{rg} = \left[1 - \frac{S_L - S_{LR}}{S_m - S_{LR}}\right]^2 \left[1 - \frac{S_L - S_{LR}}{1 - S_{LR}}\right]^2 \tag{21}$$

$$k_{ro} = \left[\frac{S_L - S_{LR}}{1 - S_{LR}}\right]^4 \tag{22}$$

$$\frac{k_{rg}}{k_{ro}} = \frac{\left[1 - \frac{S_L - S_{LR}}{S_m - S_{LR}}\right]^2 \left[1 - \frac{S_L - S_{LR}}{1 - S_{LR}}\right]^2}{\left[\frac{S_L - S_{LR}}{1 - S_{LR}}\right]^4} \tag{23}$$

where S_L is the total liquid saturation and equal to $(1 - S_g)$; S_m is the lowest oil saturation (fraction) at which the gas phase is discontinuous; and S_{LR} is the residual liquid saturation expressed as a fraction.

Corey and Rathjens[20] studied the effect of permeability variation in porous media on the value of the S_m factor in Corey's equations. They confirmed that S_m is essentially equal to unity for uniform and isotropic porous media; however, values of S_m were found to be greater than unity when there was stratification perpendicular to the direction of flow and less than unity in the presence of stratification parallel to the direction of flow. They also concluded that oil relative permeabilities were less sensitive to stratification than the gas relative permeabilities.

The gas-oil relative permeability equation is often used for testing, extrapolation, and smoothing experimental data. It is also a convenient expression that may be used in computer simulation of reservoir performance.

Corey's gas-oil relative permeability ratio equation can be solved if only two points on the k_{rg}/k_{ro} vs. S_g curve are available. However, the algebraic solution of the k_{rg}/k_{ro} equation when two points are available is very tedious and the graphical solution that Corey offers in his original paper requires lengthy graphical construction as well as numerical computation. Johnson[21] has offered a greatly simplified and useful method for determination of Corey's constant.

Johnson constructed three plots by assuming values of S_{LR}, S_m, and k_{rg}/k_{ro} by calculating the gas saturation, $(1 - S_L)$, using Corey's equations. The calculation was carried out for various S_{LR} and S_m combinations and for k_{rg}/k_{ro} values of 10 to 0.1, 1.0 to 0.01, and 0.1 to 0.001. Johnson's graphs may be used to plot a more complete k_{rg}/k_{ro} curve based on limited experimental data. The span of the experimental data determines which of the three figures should be selected.

The suggested procedure for k_{rg}/k_{ro} calculation, based on Corey's equation, is as follows:

1. Plot the experimental k_{rg}/k_{ro} vs. S_g on semilog paper with k_{rg}/k_{ro} on the logarithmic scale.
2. From the experimental data determine the gas saturation at k_{rg}/k_{ro} equal to 10.0 and 0.1, 1.0 and 0.01, or 0.1 and 0.001. (The listed pairs of values correspond to Figures 4, 5, and 6 of Johnson's data, respectively, and the range of the experimental data dictates which figure is to be employed. Note that if the data do not span the entire permeability ratio interval of 10.0 to 1.0, Figure 4 may not be employed first; instead Figure 5 with the k_{rg}/k_{ro} interval of 1.0 to 0.01 or Figure 6 with the k_{rg}/k_{ro} interval of 0.10 to 0.001 may be used first.)
3. Enter the appropriate Figure (4, 5, or 6) using the gas saturations corresponding to the pair of k_{rg}/k_{ro} values selected in step 2.
4. Pick a unique S_{LR} and S_m at the intersection of the gas saturation values; interpolate if necessary.
5. Using these S_{LR} and S_m values and employing the two other figures of Johnson, determine two more gas saturation values and the k_{rg}/k_{ro} ratio indicated on the axes of each figure.
6. Add these points to the experimental plot for obtaining the relative permeability ratio over the region of interest.

This procedure provides values of gas saturation at k_{rg}/k_{ro} ratios of 10.0, 1.0, 0.10, 0.01, and 0.001, which are sufficient to plot an expanded k_{rg}/k_{ro} curve.

It should be noted that if the data cover a wide range of permeability ratios, multiple determinations of S_{LR} and S_m can be made. If the calculated values differ from the experimental data, the discrepancy indicates that there is no single Corey curve which will fit all

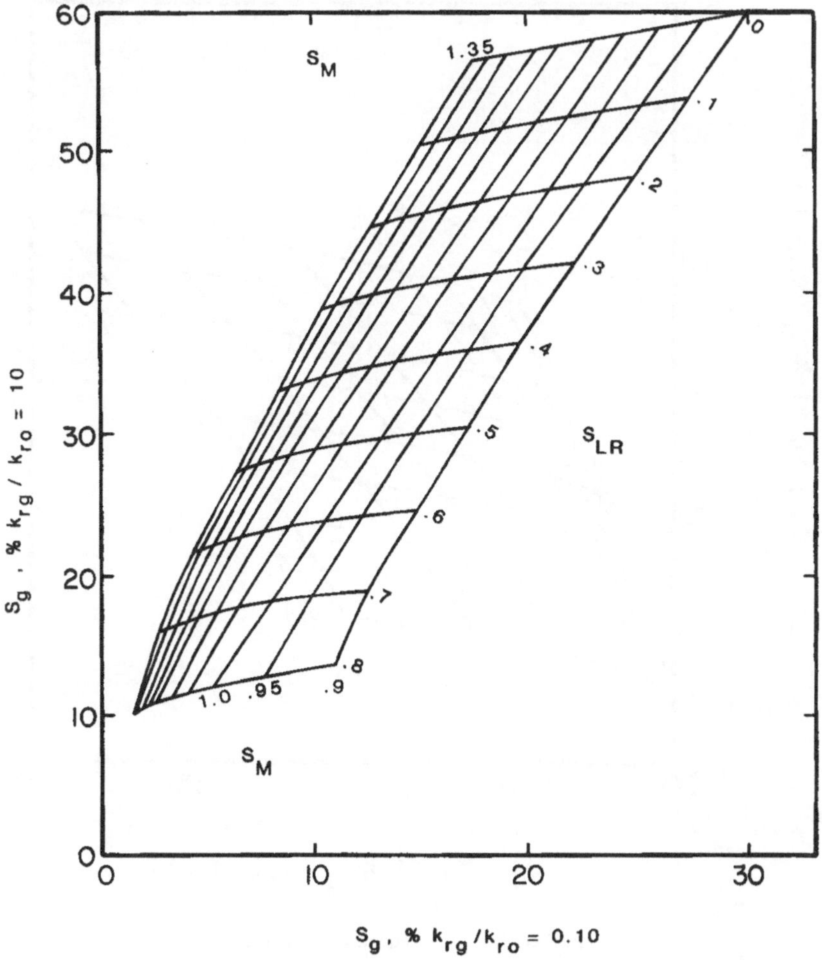

FIGURE 4. Corey equation constants.[21]

the points; an average of the values for each constant should yield a better curve fit. Figure 7 illustrates the graphical technique of Johnson.

Corey's equations for drainage oil and gas relative permeabilities and the gas-oil relative permeability ratio in the simplest form are as follows:

$$k_{ro} = (S_{oe})^4 \tag{24}$$

$$k_{rg} = (1 - S_{oe})^2 \times (1 - S_{oe}^2) \tag{25}$$

and they are related through

$$\frac{k_{ro}}{(S_{oe})^2} + \frac{k_{rg}}{(1 - S_{oe})^2} = 1 \tag{26}$$

where S_{or} represents the lowest oil saturation at which the gas tortuosity is infinite; S_{oe} is defined as $(S_o - S_{or})/(1 - S_{or})$.

Corey's equations in the presence of irreducible water saturation take the following form:

$$k_{ro} = (S^*)^4 \tag{27}$$

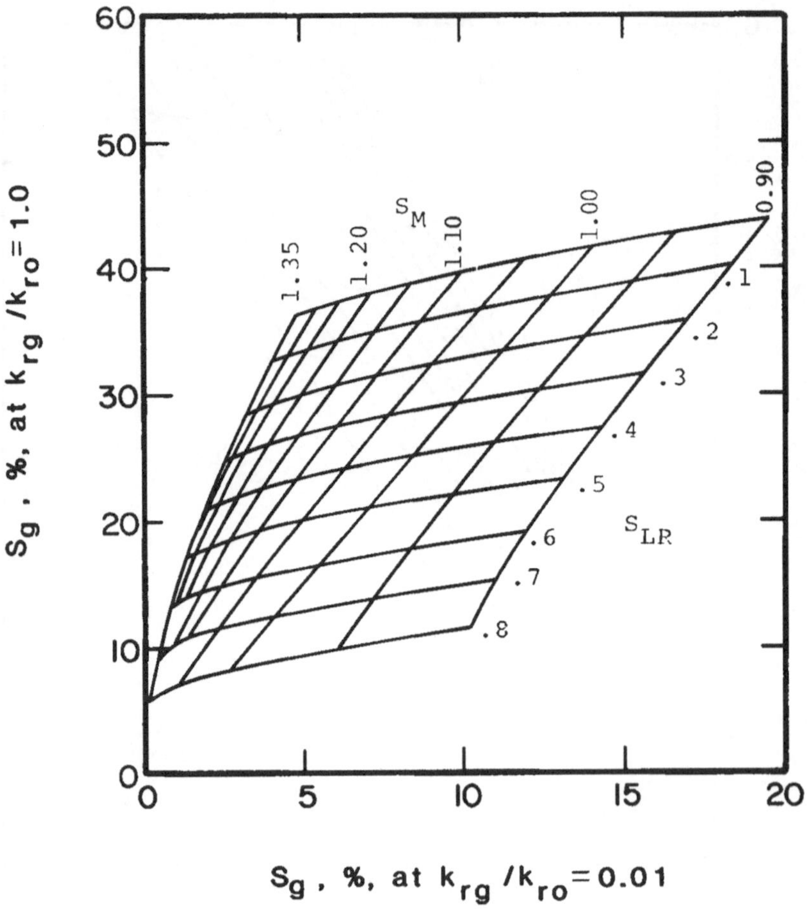

FIGURE 5. Corey equation constants.[21]

$$k_{rg} = \left[1 - \frac{S_o}{S_m - S_{wi}}\right]^2 \times (1 - S^*)^2 \tag{28}$$

where S_m is a constant related to $(1 - S_{gc})$ and as a first approximation S_m can be assumed to be unity. This is a good approximation, since S_{gc} is less than 5% in rocks with intergranular porosity. In these equations, $S^* = S_o/(1 - S_{wi})$ and S_o is the oil saturation represented as a fraction of the pore volume of the rock; S_{wi} is the irreducible water saturation, also expressed as a fraction of the pore volume.

These equations are linked by the relationship

$$\frac{k_{ro}}{(S^*)^2} + \frac{k_{rg}}{(1 - S^*)^2} = 1 \tag{29}$$

Corey et al. plotted several hundred capillary pressure-saturation curves for consolidated rocks and only a few of them met the linear relationship requirement. However, comparison of Corey's predicted relative permeabilities with experimental values for a large number of samples showed close agreement, indicating that Corey's predicted relative permeabilities are not very sensitive to the shape of the capillary pressure curves.

Equation 24 may be employed to calculate water relative permeability if the oil saturation and the residual oil saturation are replaced by water saturation and irreducible water satu-

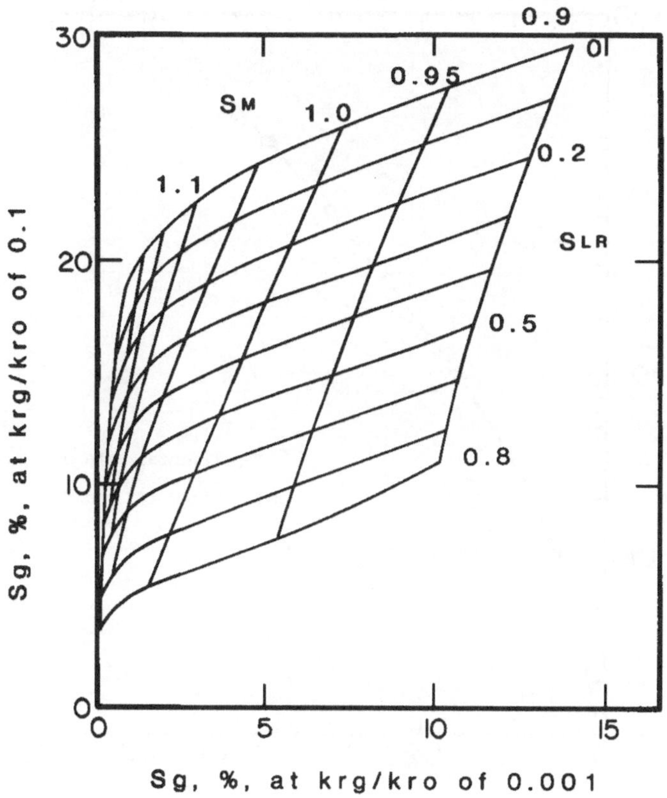

FIGURE 6. Corey equation constants.[21]

ration, respectively. The exponent of Corey's water relative permeability equation is approximately four for consolidated rocks, but depends somewhat on the size and arrangement of the pores. The exponent has a value of three for rocks with perfectly uniform pore size distribution. Several other authors have proposed similar water relative permeability equations with different exponents for other types of porous media. Values of 3.0[22] and 3.5[23] were proposed for unconsolidated sands with a single grain structure which may not be absolutely uniform in pore size but should have a narrow range of pore sizes.

Corey compared the calculated values of oil and gas relative permeabilities for poorly consolidated sands with laboratory-measured values and obtained good results. However, his results showed some deviation at low gas saturations for consolidated sandstone. Corey concluded that the equations are not valid when stratification, solution channels, fractures, or extensive consolidation is present.

Application of Corey's equation permits oil relative permeability to be calculated from measurements of gas relative permeability. Since k_{rg} measurements are easily made while k_{ro} measurements are made with difficulty, Corey's equation is quite useful. The procedure involves the measurement of gas relative permeability at several values of gas saturation in an oil-gas system and then performing the following steps:

1. Prepare an accurate plot of the function $k_{rg} = (1 - S_{oe})^2 \times (1 - S_{oe}^2)$ by assuming arbitrary values of S_{oe}, the effective saturation, which is defined as

$$S_{oe} = \frac{(S_o - S_{or})}{(1 - S_{or})}$$

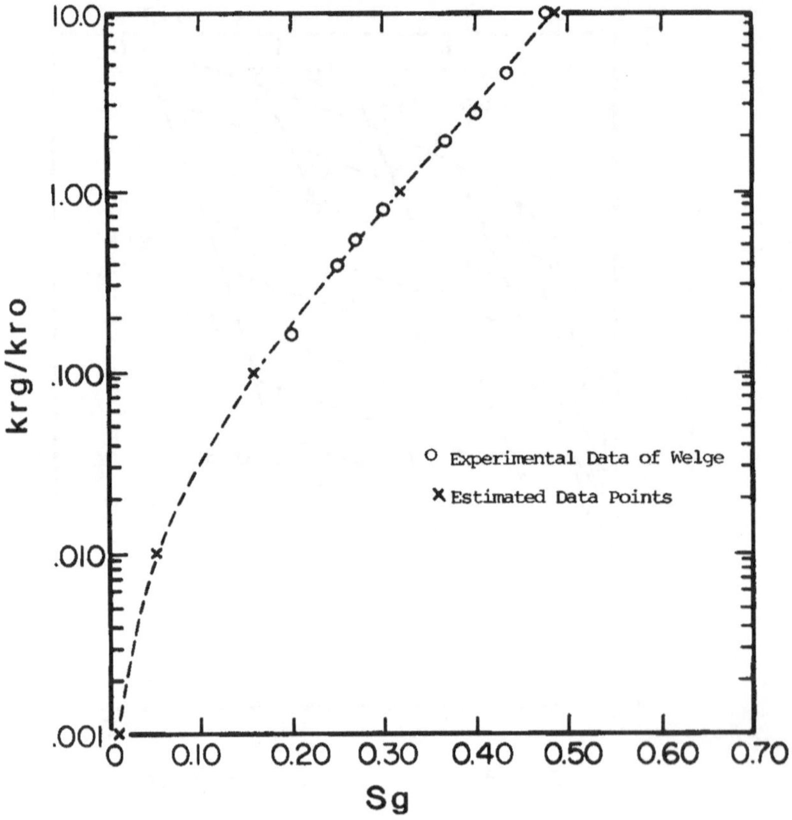

FIGURE 7. Example of the use of the Corey equations.[21]

2. Prepare a tabulation of k_{rg} vs. S_{oe} for values of k_{rg} ranging from 0.001 to 0.99 in stepwise fashion.
3. Determine values of S_{oe} for each experimental value of k_{rg} by using the above-described tabulation.
4. Plot these values of S_{oe} against the values of S_o corresponding to the k_{rg} values on rectangular coordinate paper. The plot should be a straight line between 50 and 80% oil saturation.
5. Construct a straight line through the points in this range and extrapolate to $S_{oe} = 0$. The value of S_o at this point corresponds to S_{or}. (See Figure 8.)
6. Employ Equation 24, $k_{ro} = (S_{oe})^4$ and the value of S_{or} obtained in the previous step to calculate k_{ro} values for assumed values of S_o.

Corey-type equations for drainage gas-oil relative permeability (gas drive) in the presence of connate water saturation have been suggested as follows:

$$k_{ro} = (1 - S)^4 \tag{30}$$

$$k_{rg} = S^3(2 - S) \tag{31}$$

where S represents $(S_g)/(1 - S_{wi})$.

Corey's equations for the drainage cycle in water-wet sandstones as well as carbonate formations are as follows:

$$k_{ro} = \left[\frac{1 - S_w}{1 - S_{wi}}\right]^4 \tag{32}$$

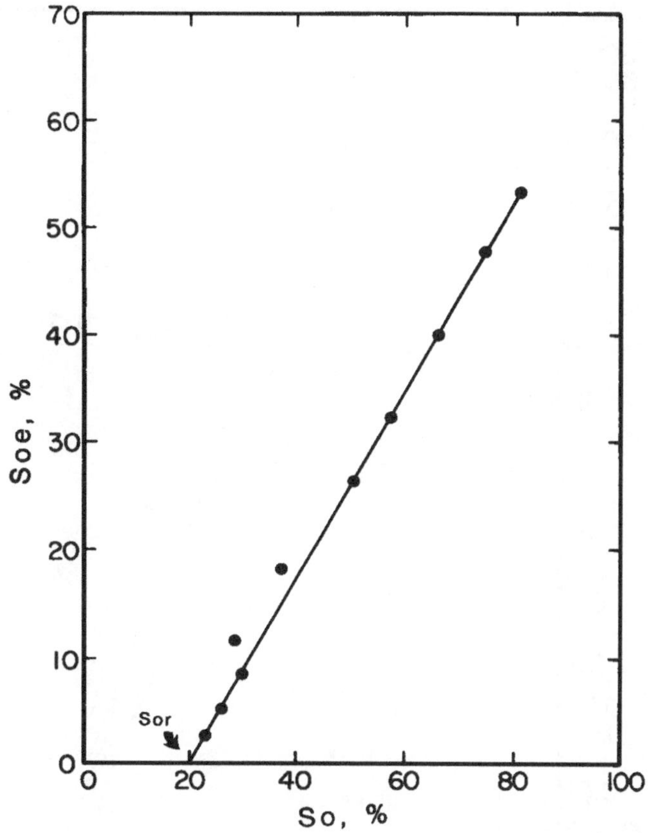

FIGURE 8. Determination of residual oil saturation based on effective oil saturation.

$$k_{rw} = (S_w^*)^4 \tag{33}$$

VII. WAHL, TORCASO, AND WYLLIE

Wahl et al.[24] suggested the use of the following equation for drainage gas-oil relative permeability ratios based on field measurements of sandstone reservoirs:

$$\frac{k_{rg}}{k_{ro}} = \psi(0.0435 + 0.4556\ \psi) \tag{34}$$

where ψ represents $(1 - S_{wc} - S_o - S_{gc})/(S_o - C)$; S_{gc} is the critical gas saturation as a fraction of total pore space; and C is a constant equal to 0.25.

Torcaso and Wyllie[25] compared gas-oil relative permeability ratios calculated by Corey's equation with those obtained from Wahl et al. for various irreducible water saturations. This comparison suggested that Corey's work was theoretically sound, since it agreed with values obtained from field measurements by Wahl et al. (see Figure 9).[24]

VIII. BROOKS AND COREY

Brooks and Corey[26,27] modified Corey's original drainage capillary pressure-saturation relationship and combined the modified equation with Burdine's equation to develop the following expression that predicts drainage relative permeability for any pore size distribution:

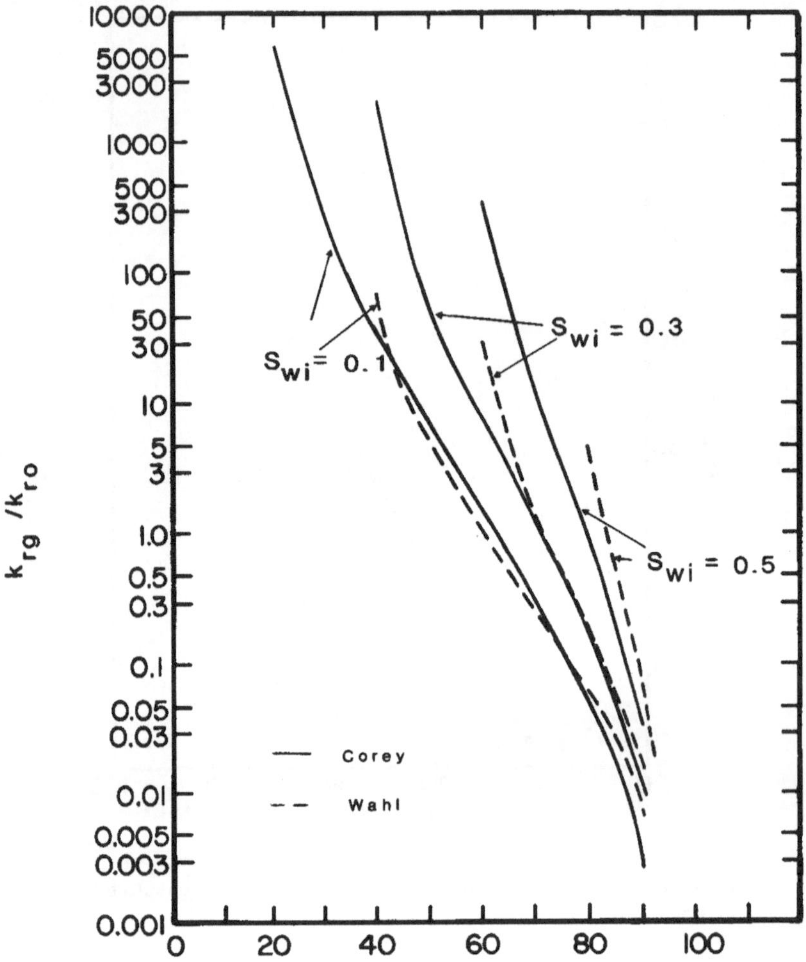

FIGURE 9. Comparison of relative permeability calculations at three irreducible water saturations.[25]

$$S_w^* = \left(\frac{P_b}{P_c}\right)^{\lambda} \text{ for } P_c \geqq P_b \tag{35}$$

where λ and P_b are constants characteristic of the media; λ is a measure of pore size distribution of the media, and P_b is a measure of maximum pore size (minimum drainage capillary pressure at which a continuous nonwetting phase exists). Using this relationship, two-phase relative permeabilities are given by

$$k_{rwt} = (S_w^*)^{\frac{2 + 3\lambda}{\lambda}} \tag{36}$$

and

$$k_{rnwt} = (1 - S_w^*)^2 \left[1 - (S_w^*)^{\frac{2 + \lambda}{\lambda}}\right] \tag{37}$$

where k_{rwt} and k_{rnwt} are wetting and nonwetting phase relative permeabilities respectively. The values of λ and P_b are obtained by plotting $(S_w - S_{wi})/(1 - S_{wi})$ vs. capillary pressure on a log-log scale and establishing a straight line with λ as the slope and P_b as the intercept at $(S_w - S_{wi})/(1 - S_{wi}) = 1$.

These equations reduce to Equations 24 and 25 for $\lambda = 2$. Theoretically λ may have any value greater than zero, being large for media with relative uniformity and small for media with wide pore size variation. The commonly encountered range for λ is between two and four for various sandstones.[27] Talash[28] obtained similar equations with somewhat different exponents.

IX. WYLLIE, GARDNER, AND TORCASO

Wyllie and Gardner[17] have presented the following expressions for the drainage water-oil relative permeability:

$$k_{rw} = \left(\frac{S_w - S_{wi}}{1 - S_{wi}}\right)^2 \frac{\int_{S_{wi}}^{S_w} dS_w/P_c^2}{\int_{S_{wi}}^{1} dS_w/P_c^2} \tag{38}$$

$$k_{ro} = \left(\frac{1 - S_w}{1 - S_{wi}}\right)^2 \frac{\int_{S_w}^{1} dS_w/P_c^2}{\int_{S_{wi}}^{1} dS_w/(P_c)^2} \tag{39}$$

More general expressions for any wetting and nonwetting relative permeability may be written where

k_{rwt} = Relative permeability to wetting phase (k_{rw} and k_{ro}).
k_{rnwt} = Nonwetting phase relative permeability (k_{rg}).
S_{wi} = Irreducible water saturation.
S_L = Total liquid saturation = $(1 - S_g)$.

$$k_{rwt} = \left(\frac{S_L - S_{wi}}{1 - S_{wi}}\right)^2 \frac{\int_{S_{wi}}^{S_L} dS_w/P_c^2}{\int_{S_{wi}}^{1} dS_w/P_c^2} \tag{40}$$

$$k_{rnwt} = \left(\frac{1.0 - S_L}{1.0 - S_{wi}}\right)^2 \frac{\int_{S_L}^{1} dS_w/P_c^2}{\int_{S_{wi}}^{1} dS_w/P_c^2} \tag{41}$$

Wyllie and Gardner have also suggested the following equation for relative permeability to water or oil when one relative permeability is available:

$$k_{rw} = (S_w^*)^2 - k_{ro}\,(S_w^*/(1 - S_w^*))^2 \tag{42}$$

where S_w^*, which is defined as $(S_w - S_{wi})/(1 - S_{wi})$, is the mobile wetting-phase saturation in a water-wet system.

Based on the linear relation between $1/P_c^2$ and $S_o/(1 - S_{wi})$, they obtained a drainage water relative permeability equation for water-wet rocks with intergranular porosity as follows:

$$k_{rw} = (S_w{}^*)^4 \tag{43}$$

Torcaso and Wyllie[25] suggested the following equation for calculation of gas-oil relative permeability of water-wet sandstone, where $1/P_c{}^2$ is approximately a linear function of effective saturation. Their derivation was based on the relation developed by Corey:

$$\frac{k_{rg}}{k_{ro}} = \frac{(1 - S^*)^2 (1 - S^{*2})}{(S^*)^4} \tag{44}$$

where S^* represents effective oil saturation and is equal to $S_o/(1 - S_{wi})$. Obviously, a reliable value of irreducible water saturation, S_{wi}, needs to be known to calculate the gas-oil relative permeability ratio.

X. LAND, WYLLIE, ROSE, PIRSON, AND BOATMAN

Land[29] reported that an appreciable adjustment of experimental parameters was required to avoid a discrepancy between experimental and calculated two-phase relative permeabilities. A large number of the relative permeability prediction methods are based on derivation of pore size distribution factors from the saturation and drainage capillary pressure relationship. Some authors[30] believe that the employment of capillary pressure relationships for the prediction of relative permeability is not advisable, since capillary pressure is derived from experiments performed under static conditions, whereas relative permeability is a dynamic phenomenon. McCaffery[31] in his thesis argues that the surface or capillary forces are orders of magnitude larger than forces arising from the fluid flow and thus, predominate in controlling the microscopic distribution of the fluid phases in many oil reservoir situations. Brown's[32] results from the measurement of capillary pressure under static and dynamic conditions support McCaffery's argument.

Several relative permeability prediction methods which are based on drainage capillary pressure curves assume that pore size distribution can be derived from these curves. These proposed models can only be applied when a strong wetting preference is known to exist.

Additionally, relative permeability calculations from capillary pressure data are developed for a capillary drainage situation where a nonwetting phase, such as gas, displaces a wetting phase (oil in a gas-oil system, or water in a gas-water system). They are developed primarily for gas-oil or gas-condensate relative permeability calculations; however, water-oil relative permeability can be calculated with a lesser certainty.

Wyllie in Frick's *Petroleum Production Handbook*[33] suggested simple empirical gas-oil and water-oil relative permeability equations for drainage in consolidated and unconsolidated sands as well as oolitic limestone rocks. These equations are presented in Tables 2 and 3.

The oil-gas and water-oil relative permeability relations for various types of rocks presented in Tables 2 and 3 may be used to produce k_{rg}/k_{ro} curves at various S_{wi} when k_{rg} measurements are unavailable.

It should be noted that the k_{ro}/k_{rw} values obtained apply only if water is the wetting phase and is decreasing from an initial value of unity by increasing the oil saturation. This is contrary to what happens during natural water drive or waterflooding processes; however, Figures 10 through 14 also apply to preferentially oil-wet systems on the drainage cycle with respect to oil if the curves were simply relabeled.

Rose[6] developed a useful method of calculating a relative permeability relationship on the basis of analysis of the physical interrelationship between the fluid flow phenomena in porous media and the static and residual saturation values. The equations for the wetting and nonwetting relative permeabilites are

Table 2
OIL-GAS RELATIVE PERMEABILITIES (FOR DRAINAGE CYCLE RELATIVE TO OIL)[33]

Type of formation	k_{ro}	k_{rg}
Unconsolidated sand, well sorted	$(S^*)^3$	$(1 - S^*)^3$
Unconsolidated sand, poorly sorted	$(S^*)^{3.5}$	$(1 - S^*)^2\,(1 - S^{*1.5})$
Cemented sandstone, oolitic limestone, rocks with vugular porosity[a]	$(S^*)^4$	$(1 - S^*)^2\,(1 - S^{*2})$

Note: In these relations the quantity $S^* = S_o/(1 - S_{wi})$.

Application to vugular rocks is possible only when the size of the vugs is small by comparison with the size of the rock unit for which the calculation is made. The unit should be at least a thousandfold larger than a typical vug.

Table 3
WATER-OIL RELATIVE PERMEABILITIES (FOR DRAINAGE CYCLE RELATIVE TO WATER)[33]

Type of formation	k_{ro}	k_{rw}
Unconsolidated sand, well sorted	$(1 - S_w^*)^3$	$(S_w^*)^3$
Unconsolidated sand, poorly sorted	$(1 - S_w^*)^2\,(1 - S_w^{*1.5})$	$(S_w^*)^{3.5}$
Cemented sandstone, oolitic limestone	$(1 - S_w^*)^2\,(1 - S_w^{*2})$	$(S_w^*)^4$

Note: In these relations the quantity $S_w^* = (S_w - S_{wi})/(1 - S_{wi})$, where S_{wi} is the irreducible water saturation.

$$k_{rw} = \frac{16S_w^2(S_w - S_{wm})^3(1 - S_{wm})}{[2S_w^2(2 - 3S_{wm}) + 3S_wS_{wm}(3S_{wm} - 2) + S_{wm}(4 - 5S_{wm})]^2} \tag{45}$$

$$k_{rn} = \frac{16S_{nwt}^2(S_{nwt} - S_{nm})^3(1 - \psi_w - S_{nm})}{[2S_{nwt}^2(2 - 2\psi_w - 3S_{nm}) + 3S_{nwt}\,S_{nm}(3S_{nm} - 2 + 2\psi_w) + S_{nm}(1 - \psi_w)(4 - \psi_w - 5S_{nm})]^2} \tag{46}$$

where S_w and S_{nwt} represent wetting and nonwetting saturations, respectively, expressed as fractions; S_{wm} and S_{nm} represent minimum wetting and nonwetting saturation values attained under dynamic flow conditions, expressed as fractions; they are the dynamic equivalents of S_{wi} and S_{or} obtained from static tests. The symbol ψ_w represents an immobile wetting-phase saturation expressed as a fraction. It is that part of the wetting-phase saturation which does not interfere with the nonwetting phase mobility and it is the maximum wetting-phase saturation at which the nonwetting relative permeability is unity. Note that Equation 46 reduces to Equation 45 for $\psi_w = 0$. The minimum wetting saturation, S_{wm}, depends on flow conditions and may be obtained by the Brownell and Katz[34] relationship of $S_{wm} = (1/86.3)\,[k/(g\,\sigma\,\cos\,\theta)\,dP/dx]^{-0.264}$ where g is the acceleration due to gravity, σ is the interfacial tension, θ is the contact angle, k is the permeability, and dP/dx is the pressure gradient.

The principal disadvantage of Rose's method is that the residual saturation of both phases must be known fairly accurately.

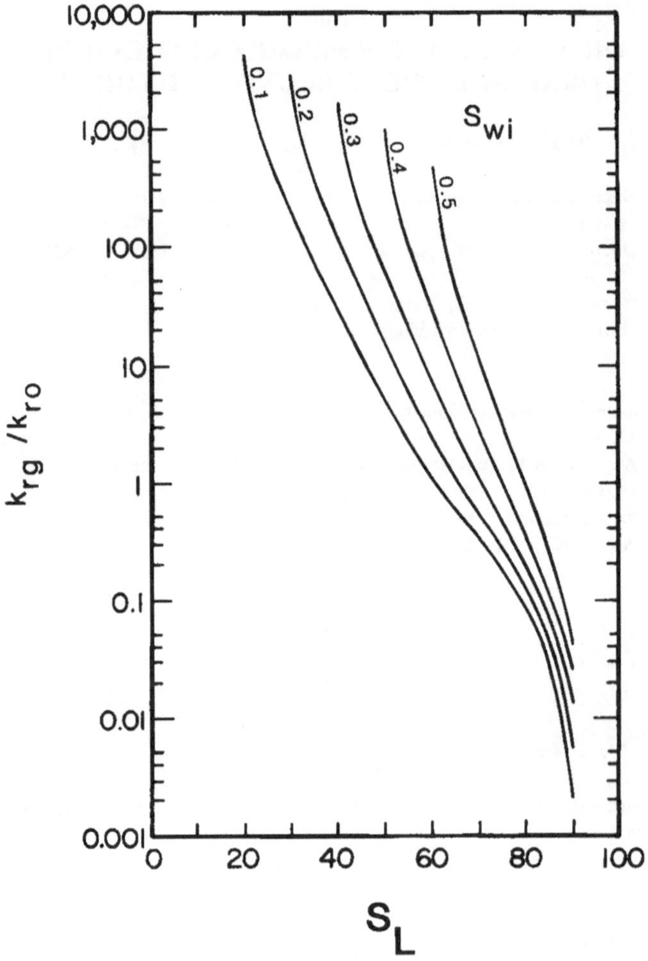

FIGURE 10. Wyllie curves for water-wet cemented sandstones, oolitic limestones, or vugular systems.[33]

Pirson[35] derived equations from petrophysical considerations for the wetting and non-wetting phase relative permeabilities in clean, water-wet, granular rocks for both drainage and imbibition processes. The water relative permeability for the imbibition cycle was given as

$$k_{rwt} = (S_w^*)^{1/2} \quad (R_o/R_t)^{3/2} \tag{47}$$

later modified to

$$k_{rwt} = (S_w^*)^{3/2} \quad (R_o/R_t)^{3/2} \tag{48}$$

and

$$k_{rwt} = (S_w^*)^{1/2} \quad S_w^4 \tag{49}$$

Water relative permeability for the drainage cycle was given by

$$k_{rwt} = (S_w^*)^{1/2} \quad S_w^3 \tag{50}$$

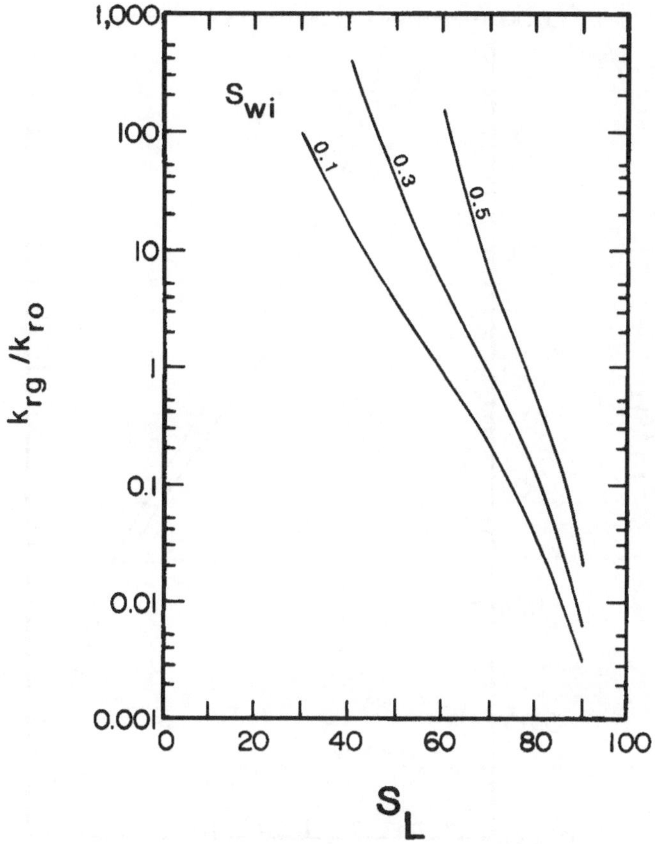

FIGURE 11. Wyllie curves for poorly sorted water-wet unconsolidated sands.[33]

where R_o represents electrical resistivity of the test core at 100% brine saturation expressed as ohm-meters; R_t represents electrical resistivity of the test core expressed as ohm-meters; S_{wi} represents irreducible wetting-phase saturation; and S_w represents water saturation as a fraction of pore space.

The nonwetting phase relative permeability in clean, water-wet rocks for the drainage cycle was found to be

$$k_{rnwt} = (1 - S_w^*)\,[1 - S_w^{*1/4}(R_o/R_t)^{1/4}]^2 \tag{51}$$

or

$$k_{rnwt} = (1 - S_w^*)\,(1 - S_w^{*1/4}\,S_w^{1/2})^2 \tag{52}$$

which was later modified to

$$k_{rnwt} = (1 - S_w^*)(1 - S_w^{*1/4}\,S_w^{1/2})^{1/2} \tag{53}$$

The nonwetting phase relative permeability in an imbibition cycle given by

$$k_{rnwt} = \left[1 - \frac{S_w - S_{wi}}{1 - S_{wi} - S_{nwt}}\right]^2 \tag{54}$$

where S_w^* represents $(S_w - S_{wi})/(1 - S_{wi})$ and S_{nwt} represents the irreducible nonwetting phase saturation as a fraction of pore space. Pirson also derived equations for the wetting

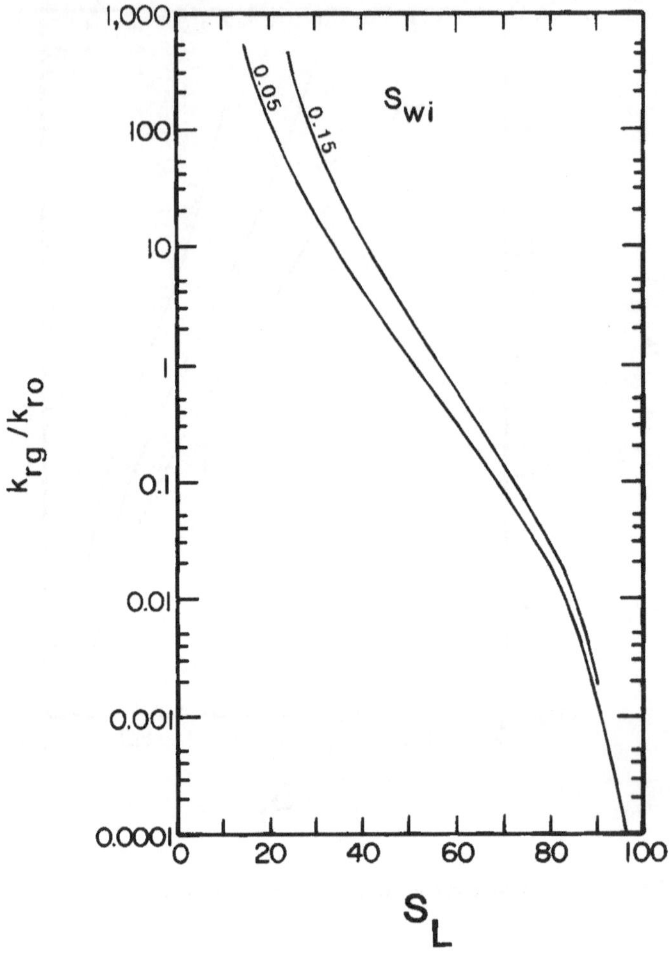

FIGURE 12. Wyllie curves for well-sorted water-wet unconsolidated cores.[33]

and nonwetting phase relative permeabilities in clean, oil-wet rocks for both drainage and imbibition processes:

$$k_{ro} = (S_{oe})^{1/2} S_o^3 \tag{55}$$

where S_{oe} is defined as $(S_o - S_{or})/(1 - S_{or})$ and S_{or} represents irreducible oil saturation and is the equilvalent of of $(1 - S_{wi})$ for a clean, water-wet rock; S_o represents total oil saturation obtained by differences from $(1 - S_w)$.

The nonwetting phase relative permeability in clean, oil-wet rocks for the imbibition cycle was found to be

$$k_{rnwt} = \left[1 - \frac{S_o - S_{or}}{1 - S_{or} - S_{wt}}\right]^2 \tag{56}$$

and for the drainage cycle was found to be

$$k_{rnwt} = (1 - S_{oe})\,[1 - S_{oe}^{1/4}\,S_o^{1/2}]^2 \tag{57}$$

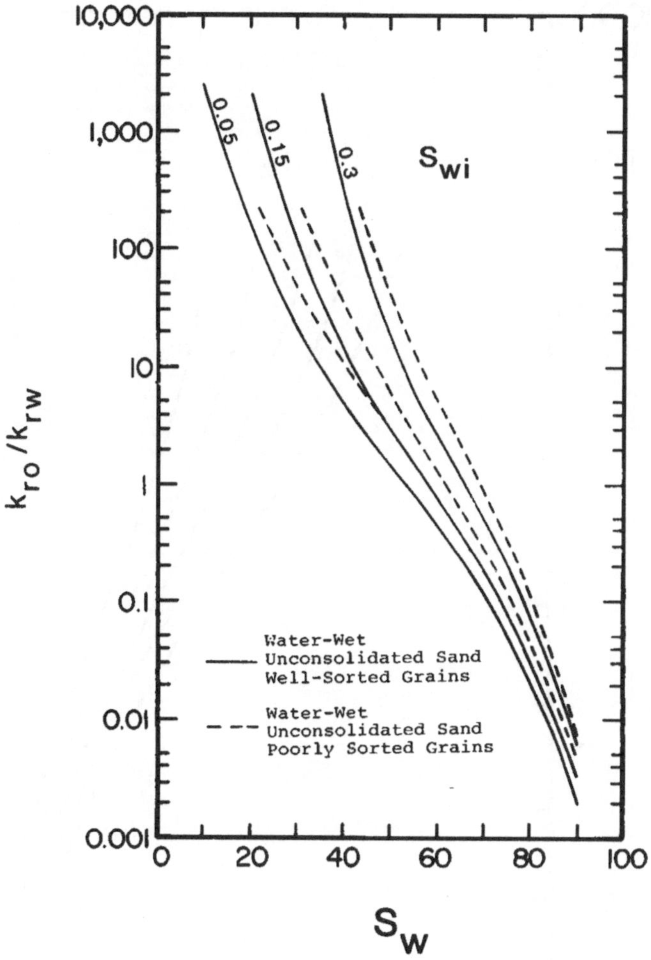

FIGURE 13. Wyllie curves.[33]

where S_{wt} represents trapped-water saturation, which is determinable by Albert and Butault's method.[36] These investigators suggested that a capillary-pressure curve be obtained either with a wetting fluid or with a nonwetting fluid such as mercury to obtain irreducible nonwetting phase saturation. They also estimated that the irreducible nonwetting phase saturation is two thirds of net pore volume made up of capillaries of radii smaller than the most common capillary size, when the nonwetting phase displaces the wetting phase.

Pirson suggested a method to determine the *in situ* trapped nonwetting phase saturation by means of microresistivity logging devices, which respond to the flushed zone around a well bore:

$$S_{nwt} = 1 - (1/\phi) \quad (R_{mf}/R_{xo})^{1/2} \tag{58}$$

where ϕ represents the porosity of the reservoir rock and R_{mf}/R_{xo} is the ratio of the mud-filtrate resistivity to flushed zone resistivity.

Boatman[37] suggested water and gas relative permeability equations in terms of core petrophysical properties obtained from laboratory data:

$$k_{rw} = S_w{}^{*3/2} (R_o/R_t)^{3/2} \tag{59}$$

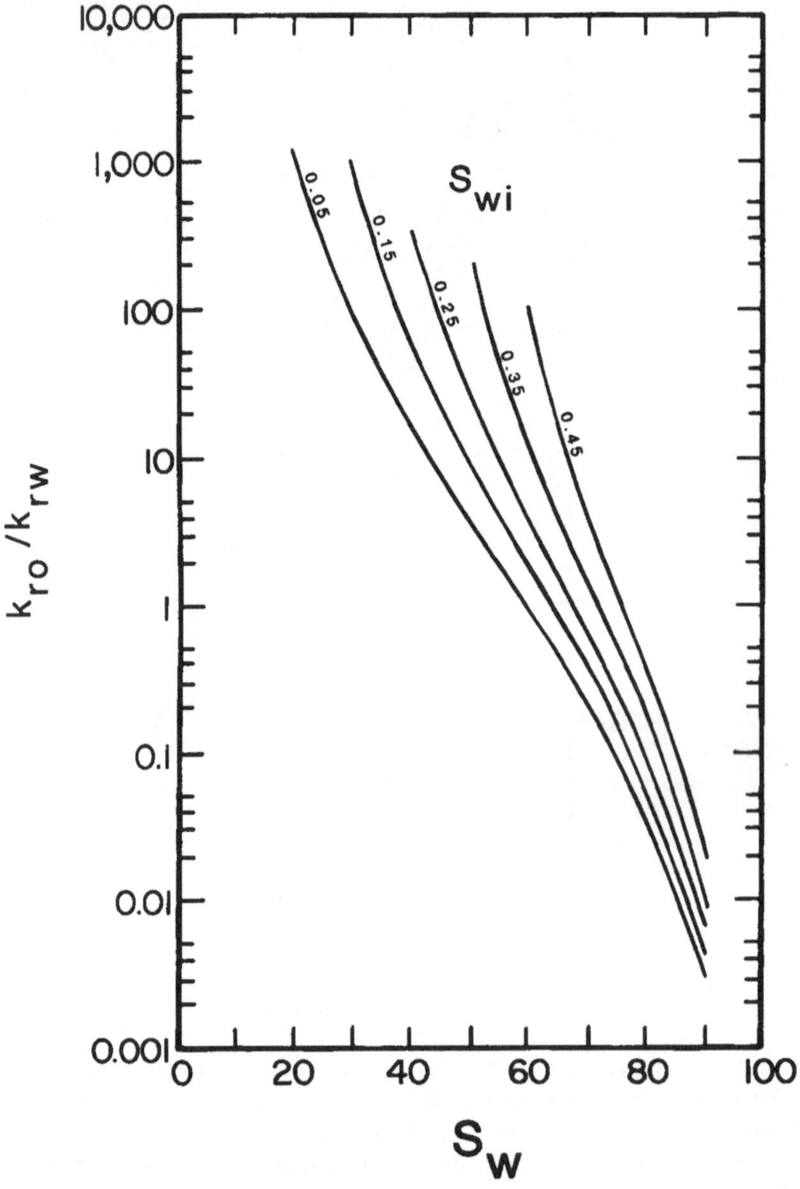

FIGURE 14. Wyllie curves for water-wet cemented sandstones, oolitic limestones, or vugular systems.[33]

and

$$k_{rg} = (1 - S_w^{*1/4}\, S_w^{1/2})^{1/2} \quad (60)$$

where

$$S_w^* = \frac{S_w - S_{wi}}{1 - S_{wi}}$$

Pirson et al.[38] proposed equations for oil and water relative permeabilities as follows:

$$k_{rw} = (S_w^*)^{1/2}\,(R_o/R_t)^2 \quad (61)$$

and

$$k_{ro} = (1 - S_{wm})^2 \tag{62}$$

where S_{wm} represents $(S_w - S_{wirr})/(1 - S_{wirr} - S_{or})$; $S_w{}^*$ represents $(S_w - S_{wirr})/(1 - S_{wirr})$.

Thornton[5] proposed the following equation for wetting-phase relative permeability:

$$k_{rwt} = S_w^3 \, (P_D/P_c)^2 \tag{63}$$

where P_D/P_c represents the ratio of displacement pressure to drainage capillary pressure.

Rose and Wyllie[7,39] proposed a petrophysical equation for wetting-phase relative permeability:

$$k_{rwt} = (I^{1/2}) \tag{64}$$

where I represents resistivity index, R_t/R_o.

Jones[40] proposed mathematical relationships for water-oil and water-gas relative permeabilities as function of S_w and S_{wi}, where S_w may be determined from well logs, S_{wi} may be estimated from an S_w - ϕ crossplot, and ϕ may be determined from well logs:

$$k_{rw} = (S_w{}^*)^3 \tag{65}$$

$$k_{ro} = \left[\frac{(0.9 - S_w)}{(0.9 - S_{wi})}\right]^2 \tag{66}$$

XI. KNOPP, HONARPOUR ET AL., AND HIRASAKI

Knopp[41] developed a correlation from 107 experimentally determined gas-oil relative permeability ratios of Venezuelan core samples. The core samples were from consolidated as well as poorly consolidated sandstone reservoirs of high porosity and permeability; the Welge gas-flood procedure was used for k_{rg}/k_{ro} determination.

A single correlation was established on the basis of the restored-state water saturation as a correlating parameter. The correlation is shown as a family of most probable k_{rg}/k_{ro} curves in Figure 15.

Comparison of Knopp's correlation with experimental values is more promising when the geometric mean of the suite of k_{rg}/k_{ro} curves for a given reservoir or sample group is compared with the corresponding most probable curves for the correlation. Knopp also suggested a procedure for developing similar correlations for various other formations.

A comparison of Knopp's correlations with the correlation of Corey and Wahl et al. on the basis of 15% water saturation is shown in Figure 16.

Honarpour et al.[42] developed a set of empirical prediction equations for water-oil imbibition relative permeability and gas-oil drainage relative permeability from a large number of experimental data. Their results are presented in Tables 4 and 5. Symbols used in these two tables are defined as follows:

k_a = air permeability, md
k_o = oil permeability, md
$k_{o(Swi)}$ = oil permeability at irreducible water saturation, md
k_{rg} = gas relative permeability, oil and gas system, fraction
$k_{rg(Sor)}$ = gas relative permeability at residual oil saturation, fraction
$k_{ro,w}$ = oil relative permeability, water and oil system, fraction
k_{rw} = water relative permeability, water and oil system, fraction
$k_{ro,g}$ = oil relative permeability, oil and gas system, fraction

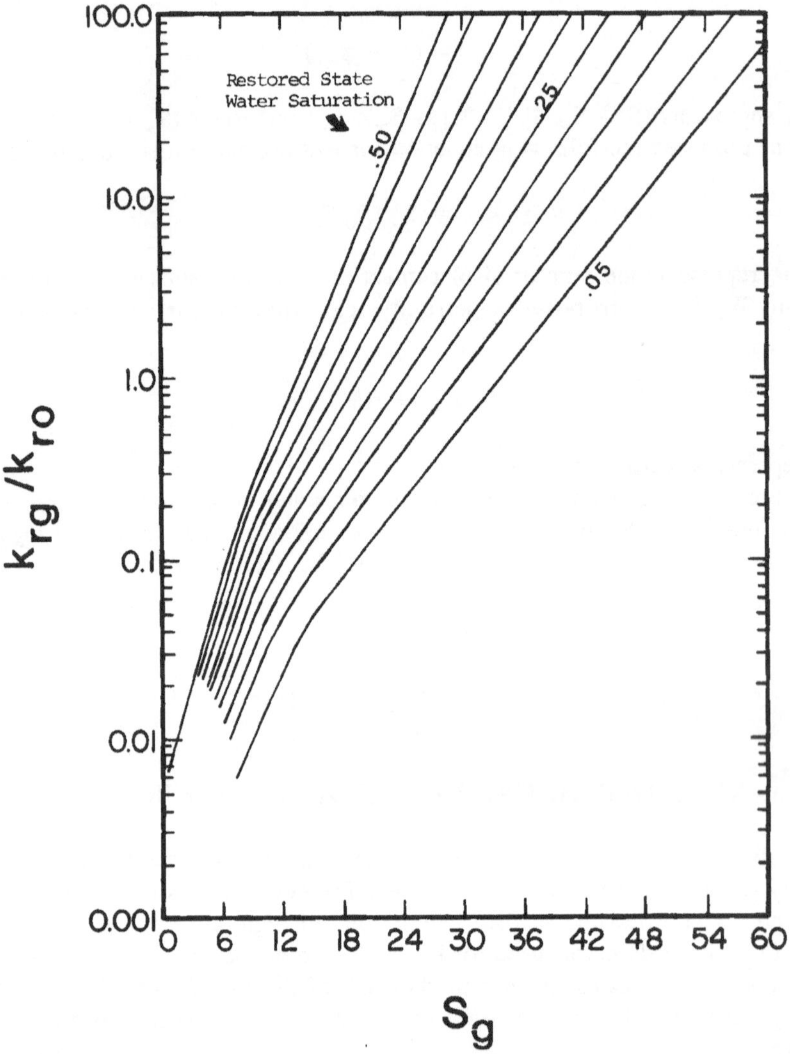

FIGURE 15. Knopp's correlation of most probable relative permeability ratios.[41]

S_g = gas saturation, fraction
S_{gc} = critical gas saturation, fraction
S_o = oil saturation, fraction
S_{org} = residual oil saturation to gas, fraction
S_{orw} = residual oil saturation to water, fraction
S_w = water saturation, fraction
S_{wi} = irreducible water saturation, fraction
ϕ = porosity, fraction

The data which were used as a basis for the study by Honarpour et al. were derived from oil and gas fields located in the continental U.S., Alaska, Canada, Libya, Iran, Argentina, and the United Arab Republic. All of the laboratory tests were made at room temperature and atmospheric pressure. No attempt was made by the authors to group the data according to laboratory techniques used in measuring relative permeability, since this information was not available for many of the data sets. Each set of relative permeability data was classified

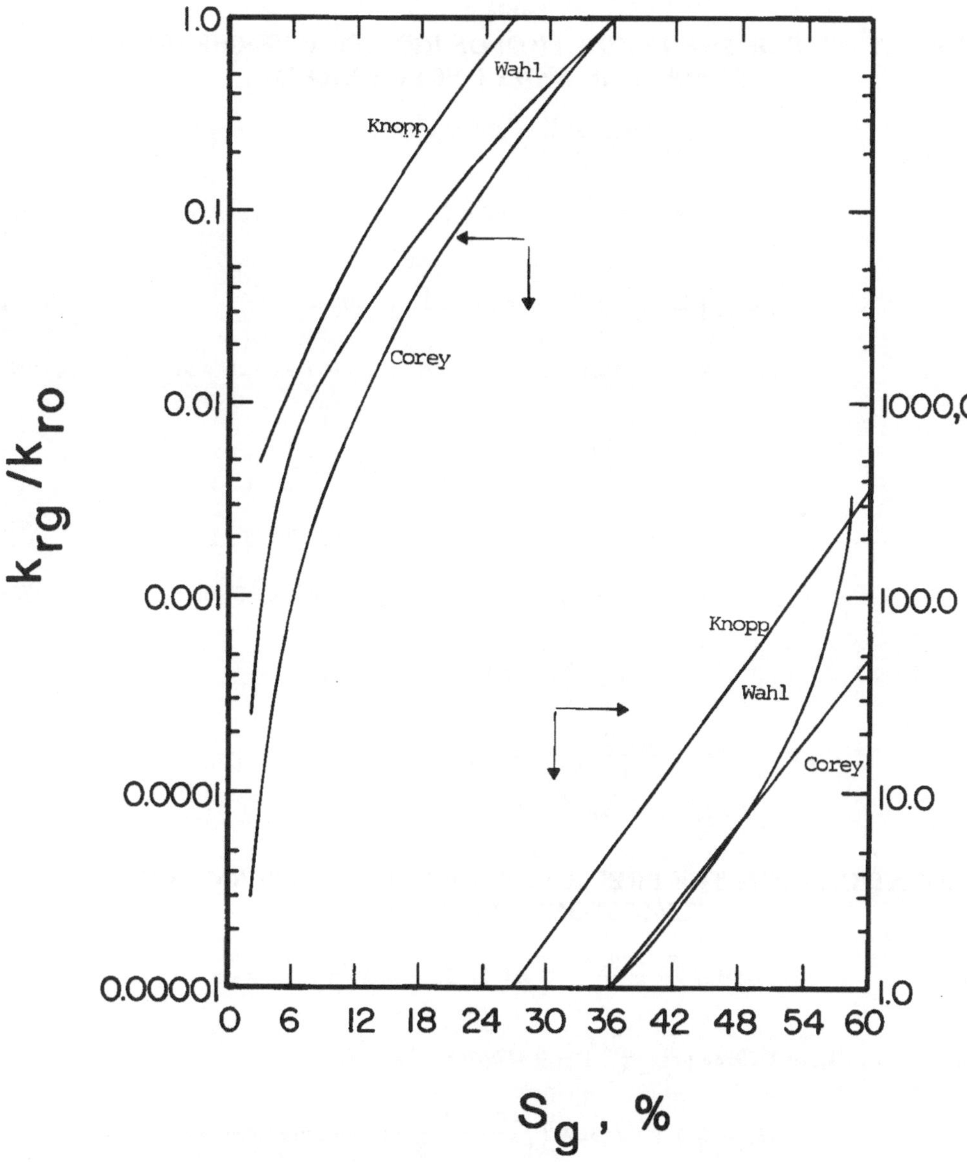

FIGURE 16. Comparison of relative permeability correlations.[41]

as either "carbonate" or "noncarbonate", but the information which was available was not sufficient for more detailed lithologic characterization.

In addition to the classification of data sets as "carbonate" or "noncarbonate", a further classification was made on the basis of wettability. This rough classification was made according to the following arbitrary criteria:

1. The rock was considered to be strongly water-wet if k_{ro} at high oil saturations in an oil-water system greatly exceeded k_{ro} in a gas-oil system at the same saturations, provided k_{rg} in a gas-oil system greatly exceeded k_{rw} in an oil-water system at or near residual oil saturation after water-flooding.
2. The rock was considered to be oil-wet when k_{ro} in the oil-water system was approximately equal to k_{ro} in the gas-oil system, provided k_{rg} in the gas-oil system was approximately equal to k_{rw} in the oil-water system.

Table 4
EQUATIONS FOR THE PREDICTION OF RELATIVE PERMEABILITY IN SANDSTONE AND CONGLOMERATE

$$k_{rw} = 0.035388 \frac{(S_w - S_{wi})}{(1 - S_{wi} - S_{orw})} - 0.010874*$$

$$\left[\frac{(S_w - S_{orw})}{(1 - S_{wi} - S_{orw})}\right]^{2.9} + 0.56556(S_w)^{3.6}(S_w - S_{wi}) \quad \text{(water-wet)} \qquad (67)$$

$$k_{rw} = 1.5814 \left[\frac{S_w - S_{wi}}{1 - S_{wi}}\right]^{1.91} - 0.58617 \frac{(S_w - S_{orw})}{(1 - S_{wi} - S_{orw})}*$$

$$(S_w - S_{wi}) - 1.2484\phi(1 - S_{wi})(S_w - S_{wi}) \quad \text{(intermediately wet)} \qquad (68)$$

$$k_{ro,w} = 0.76067 \left[\frac{\left(\frac{S_o}{1 - S_{wi}}\right) - S_{or}}{1 - S_{orw}}\right]^{1.8} \left[\frac{S_o - S_{orw}}{1 - S_{wi} - S_{orw}}\right]^{2.0}$$

$$+2.6318\phi(1 - S_{orw})(S_o - S_{orw}) \quad \text{(any wettability)} \qquad (69)$$

$$k_{ro,g} = 0.98372 \left(\frac{S_o}{1 - S_{wi}}\right)^4 \left[\frac{S_o - S_{org}}{1 - S_{wi} - S_{org}}\right]^2 \quad \text{(any wettability)} \qquad (70)$$

$$k_{rg} = 1.1072 \left(\frac{S_g - S_{gc}}{1 - S_{wi}}\right)^2 k_{rg(S_{org})} + 2.7794*$$

$$\frac{S_{org}(S_g - S_{gc})}{(1 - S_{wi})} k_{rg(S_{org})} \quad \text{(any wettability)} \qquad (71)$$

Table 5
EQUATIONS FOR THE PREDICTION OF RELATIVE PERMEABILITY IN LIMESTONE AND DOLOMITE

$$k_{rw} = 0.0020525 \frac{(S_w - S_{wi})}{\phi^{2.15}} - 0.051371 (S_w - S_{wi}) \left(\frac{1}{k_a}\right)^{0.43} \quad \text{(water-wet)} \qquad (72)$$

$$k_{rw} = 0.29986 \left(\frac{S_w - S_{wi}}{1 - S_{wi}}\right) - 0.32797 \left(\frac{S_w - S_{orw}}{1 - S_{wi} - S_{orw}}\right)^2*$$

$$(S_w - S_{wi}) + 0.413259 \left(\frac{S_w - S_{wi}}{1 - S_{wi} - S_{orw}}\right)^4 \quad \text{(intermediately wet)} \qquad (73)$$

$$k_{ro,w} = 1.2624 \left(\frac{S_o - S_{orw}}{1 - S_{orw}}\right) \left(\frac{S_o - S_{orw}}{1 - S_{wi} - S_{orw}}\right)^2 \quad \text{(any wettability)} \qquad (74)$$

$$k_{ro,g} = 0.93752 \left(\frac{S_o}{1 - S_{wi}}\right)^4 \left(\frac{S_o - S_{org}}{1 - S_{wi} - S_{org}}\right)^2 \quad \text{(any wettability)} \qquad (75)$$

$$k_{rg} = 1.8655 \frac{(S_g - S_{gc})(S_g)}{(1 - S_{wi})} k_{rg(S_{org})} + 8.0053*$$

$$\frac{(S_g - S_{gc})(S_{org})^2}{(1 - S_{wi})} - 0.025890 (S_g - S_{gc})*$$

$$\left(\frac{1 - S_{wi} - S_{org} - S_{gc}}{1 - S_{wi}}\right)^2*$$

$$\left(1 - \frac{1 - S_{wi} - S_{org} - S_{gc}}{1 - S_{wi}}\right)^2 \left(\frac{k_a}{\phi}\right)^{0.5} \quad \text{(any wettability)} \qquad (76)$$

3. The rock was considered to be of intermediate wettability when it did not clearly meet either the water-wet or the oil-wet classification criteria.

After the data sets had been classified according to lithology and wettability, stepwise linear regression analysis was employed to develop equations which would approximate the measured relative permeabilities from such factors as fluid saturations, permeability, and porosity.

All water-oil system equations refer to displacement of oil by water and the oil-gas system equations refer to drainage processes. All experimental data were measured in consolidated rocks.

The equations that were developed by Honarpour et al. have not yet been extensively tested. However, most of the tests which have been made indicated that the equations are in closer agreement with laboratory data than the predictions of published correlations which were used as a basis for comparison.

In using empirical relationships such as those presented by Honarpour et al., any calculated relative permeability which exceeds 1.0 should be assumed equal to 1.0. If a relative permeability value is known at any water saturation, the relative permeability curve may be shifted to match the known data point.

Hirasaki[43] has suggested a relative permeability correlation for fractured reservoirs as follows:

$$S^* = \frac{S_d - S_{de}}{1 - S_w - S_{de}} \tag{77}$$

$$k_{rd} = k_{rd}^{o} (S^*)^n \tag{78}$$

$$k_{ro} = k_{ro}^{o} (1 - S^*)^n \tag{79}$$

where

S^* = Normalized saturation.
S_d = Displacing phase saturation.
S_{de} = Immobile displacing phase saturation.
S_{or} = Residual oil saturation.
k_{rd} = Displacing phase relative permeability.
k^o_{rd} = Displacing phase relative permeability at residual oil saturation.
k_{ro} = Relative permeability to oil.
k^o_{ro} = Relative permeability to oil at immobile displacing phase saturation.
n = Exponent parameter for shape of relative permeability curves, said to be equal to one in fractured reservoirs.

REFERENCES

1. **Pullien, F. A. L., Ed.,** *Porous Media: Fluid Transport and Pore Structure*, Academic Press, New York, 1979.
2. **Kozeny, J.,** Uber Kapillare Leitung des Wassersim Boden, *Sitzungsber. Akad. Wiss. Wien. Math. Naturwiss. KL., Abt. 2A*, 136, 271, 1927.
3. **Purcell, W. R.,** Capillary pressures — their measurement using mercury and the calculation of permeability therefrom, *Trans. AIME*, 186, 39, 1949.

4. **Rose, W. D. and Bruce, W. A.,** Evaluation of capillary character in petroleum reservoir rock, *Trans. AIME*, 186, 127, 1949.
5. **Thornton, O. F.,** Valuation of relative permeability, *Trans. AIME*, 186, 328, 1949.
6. **Rose, W. D.,** Theoretical generalization leading to the evaluation of relative permeability, *Trans. AIME*, 186, 111, 1949.
7. **Rose, W. and Wyllie, M. R. J.,** Theoretical description of wetting liquid relative permeability, *Trans. AIME*, 186, 329, 1949.
8. **Gates, J. I. and Leitz, W. J.,** Relative permeabilities of California cores by the capillary pressure method, paper presented at the API Meeting, Los Angeles, California, May 11, 1950, 286.
9. **Rapoport, L. A. and Leas, W. J.,** Relative permeability to liquid in liquid-gas system, *Trans. AIME*, 192, 83, 1951.
10. **Wyllie, M. R. J.,** Interrelationship between wetting and non-wetting phase relative permeability, *Trans. AIME*, 192, 83, 1981.
11. **Fatt, I. and Dykstra, H.,** Relative permeability studies, *Trans. AIME*, 192, 249, 1951.
12. **Wyllie, M. R. J. and Sprangler, M. B.,** Application of electrical resistivity measurements to problems of fluid flow in porous media, *Bull. AAPG*, 36, 359, 1952.
13. **Burdine, N. T.,** Relative permeability calculations from pore size distribution data, *Trans. AIME*, 198, 71, 1953.
14. **Naar, J. and Henderson, J. H.,** An imbibition model — its application to flow behavior and the prediction of oil recovery, *Trans. AIME*, 222, 61, 1961.
15. **Naar, J. and Wygal, R. J.,** Three-phase imbibition relative permeability, *Trans. AIME*, 222, 254, 1961.
16. **Land, C. S.,** Calculation of imbibition relative permeability for two- and three-phase flow from rock properties, *Soc. Pet. Eng. J.*, 6, 149, 1968.
17. **Wyllie, M. R. J. and Gardner, G. H. F.,** The generalized Kozeny-Carmen equation, its application to problems of multi-phase flow in porous media, *World Oil*, 146, 121, 1958.
18. **Timmerman, E. H., Ed.,** *Practical Reservoir Engineering*, Penwell Publ., 1982, 101.
19. **Corey, A. T.,** The interrelation between gas and oil relative permeabilities, *Prod. Mon.*, 19, 38, 1954.
20. **Corey, A. T. and Rathjens, C. H.,** Effect of stratification on relative permeability, *Trans. AIME*, 207, 358, 1956.
21. **Johnson, C. E., Jr.,** Graphical determination of the constants in the Corey equation for gas-oil relative permeability ratio, *J. Pet. Technol.*, 10, 1111, 1968.
22. **Irmay, S.,** On the hydraulic conductivity of unsaturated soils, *Trans. AGU*, 35(3), 463, 1954.
23. **Averganov, S. F.,** *About Permeability of Subsurface Soils in Case of Incomplete Saturation*, Engineering Collection, Vol. 7, 1950, cited by Polubarinova-Kochina, P, in *The Theory of Ground Water Movement*, English translation by Dewiest, R. J. M., Princeton Univ. Press, Princeton, N.J., 1962.
24. **Wahl, W. L., Mullins, L. D., and Elfrink, E. B.,** Estimation of ultimate recovery from solution gas drive reservoirs, *Trans. AIME*, 213, 132, 1958.
25. **Torcaso, M. A. and Wyllie, M. R. J.,** A comparison of calculated k_{rg}/k_{ro} ratios with field data, *J. Pet. Technol.*, 6, 57, 1958.
26. **Brooks, R. H. and Corey, A. T.,** *Hydraulic Properties of Porous Media*, Hydrology Papers, No. 3, Colorado State University, Ft. Collins, Colo., 1964.
27. **Brooks, R. H. and Corey, A. T.,** Properties of porous media affecting fluid flow, *J. Irrig. Drain. Div.*, 6, 61, 1966.
28. **Talash, A. W.,** Experimental and calculated relative permeability data for systems containing tension additives, Paper 5810, *Society of Petroleum Engineers*, Dallas, Tx., 1976.
29. **Land, C. S.,** Calculation of imbibition relative permeability for two- and three-phase flow from rock properties, *Soc. Pet. Eng. J.*, 6, 149, 1968.
30. **Bear, J, Ed.,** *Dynamics of Fluids in Porous Media*, Elsevier, Amsterdam, 1972.
31. **McCaffery, F. G.,** The Effect of Wettability of Relative Permeability and Imbibition in Porous Media, Ph.D. thesis, University of Calgary, Alberta, Canada, 1973.
32. **Brown, H. W.,** Capillary pressure investigations, *Trans. AIME*, 192, 67, 1951.
33. **Frick, T., Ed.,** *Petroleum Production Handbook*, Vol. 2, Society of Petroleum Engineers of AIME, Dallas, Tx., 1962, 25.
34. **Brownell, L. E. and Katz, D.,** Flow of fluids through porous media, *Chem. Eng. Prog.*, 43(11), 603, 1947.
35. **Pirson, S. J.,** *Ed., Oil Reservoir Engineering*, McGraw Hill, New York, 1958.
36. **Albert, P. and Butault, L.,** Etude des Characteristiques Capillaries du Reservoir du Cap don Par La Methode Purcell, *Pet. Ann. Combus. Liq.*, 7(8), 250, 1952.
37. **Boatman, E. M.,** An Experimental Investigation of Some Relative Permeability-Relative Conductivity Relationships, M.S. thesis, University of Texas, Austin, 1961.
38. **Pirson, S. J., Boatman, E. M., and Nettle, R. L.,** Prediction of relative permeability characteristics of intergranular reservoir rocks from electrical resistivity measurements, *Trans. AIME*, 231, 564, 1964.

39. **Wyllie, M. R. J. and Rose, W. D.,** Some theoretical considerations related to quantitative evaluation of physical characteristics of reservoir rock from electrical log data, *Trans. AIME,* 189, 105, 1950.
40. **Jones, M. A.,** Waterflood mobility control: a case history, *J. Pet. Technol.,* 9, 1151, 1966.
41. **Knopp, C. R.,** Gas-oil relative permeability ratio correlation from laboratory data, *J. Pet. Technol.,* 9, 1111, 1965.
42. **Honarpour, M. M., Koederitz, L. F., and Harvey, A. H.,** Empirical equations for estimating two-phase relative permeability in consolidated rock, *Trans. AIME,* 273, 2905, 1982.
43. **Hirasaki, G. J.,** Estimation of Reservoir Parameters by History Matching Oil Displacement by Water or Gas, Paper 4283, Society Petroleum Engineers, Dallas, Tex., 1975.
44. **Koplik, J. and Lasseter, T. J.,** Two-phase flow in random network models of porous media, *Soc. Pet. Eng. J.,* 25, 89, 1985.
45. **Fulcher, R. A., Ertekin, T., and Stahl, C. D.,** Effect of cappillary number and its constituents on two-phase relative permeability curves, *J. Pet. Technol.,* 2, 249, 1985.

Chapter 3

FACTORS AFFECTING TWO-PHASE RELATIVE PERMEABILITY

I. INTRODUCTION

The first published information concerning the simultaneous flow of multiple fluid phases was probably by Hassler et al.[1] The term "relative permeability" had not yet been coined and Hassler et al. studied only the flow characteristics of the gas phase as a function of fluid saturation in consolidated rocks. The relative permeability concept was first postulated by Muskat and Meres.[2] Their work consisted of extending Darcy's law to two-phase systems. For oil reservoirs, the relevant two-phase fluid combinations are water-oil and liquid-gas (usually thought of as oil-gas). Gas-water relative permeability curves are used to describe the performance of gas reservoirs and gas-liquid curves are used for condensate reservoirs.

II. TWO-PHASE RELATIVE PERMEABILITY CURVES

Water-oil relative permeability is usually plotted as a function of water saturation, as shown by Figure 1. At the irreducible water saturation (S_{wc}), the water relative permeability is zero and the oil relative permeability with respect to water is some value less than one. At this point only oil can flow and the capability of the oil to flow is reduced by the presence of connate water. The effect of connate water in reducing oil flow rate is illustrated schematically by Figure 2.

Note that data to the left of the irreducible water saturation are not useful for predicting hydrocarbon reservoir performance, since water saturations less than S_{wc} are not encountered. As water saturation increases, the water relative permeability increases and the oil relative permeability (with respect to water) decreases. A maximum water saturation is reached at the residual oil saturation and the oil relative permeability becomes zero. Obviously, aquifer conditions are represented by a relative permeability to water of unity, which occurs at a water saturation of 100%.

Unfortunately, there is an alternate definition of relative permeability currently in use. This terminology (illustrated by Figure 3) defines the oil relative permeability at irreducible water saturation as having a value of one, and defines absolute permeability as the effective permeability at irreducible water saturation. The effective permeabilities are identical with both definitions of relative permeability and one set of values may be readily converted to the other. This second definition of relative permeability (k_{r2}) applies to both the oil and water phases.

These alternate or normalized values of relative permeability may be converted to standard values by

$$k_{rSTD} = k_{r2}\, k_{as}/k_{aSTD} \tag{1}$$

where

$$k_{as} = k_{eo} \text{ at } S_{wc}$$

Also note that under this second definition of relative permeability, the water relative permeability in an aquifer has a value greater than unity. Essentially, with this alternate definition, relative permeability is normalized to the value at irreducible water saturation.

Gas-oil relative permeability and gas-liquid relative permeability are similar in concept to water-oil relative permeability. The preferred relative permeability values are those taken with connate water present at the irreducible saturation value.

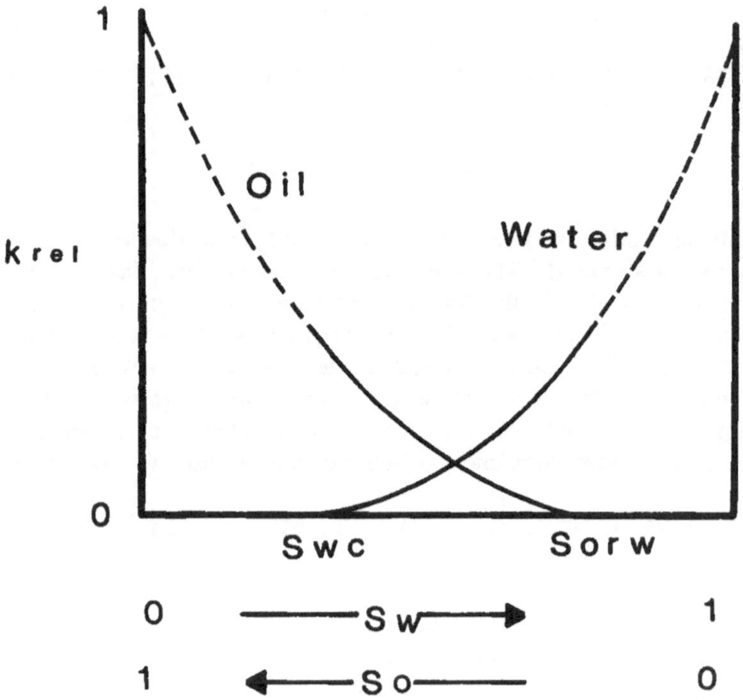

FIGURE 1. Water-oil relative permeability curves.

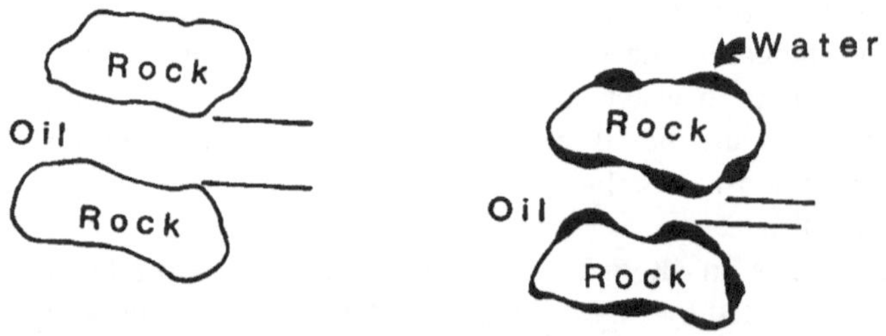

FIGURE 2. Oil flow reduction due to the presence of water.

As free gas saturation increases, the oil relative permeability with respect to gas decreases; however, until the critical gas saturation (S_{gc}) is reached, the gas relative permeability is zero. The critical gas saturation is the point at which the gas bubbles become large enough to break through the oil and away from the rock surface. As gas saturation increases, the gas relative permeability increases and theoretically reaches a value of unity at 100% gas. A gas-oil relative permeability curve is illustrated by Figure 4.

An experimental procedure to determine relative permeability in an unconsolidated sand was first described by Wyckoff and Botset.[3] Their work consisted of injecting a combination of liquids and gases through a sample under steady-state conditions. Their results are shown in Figure 5, where k_{ro} and k_{rg} are relative permeability to oil and gas, respectively. The figure is typical of wetting- and nonwetting phase relative permeabilities, regardless of whether the system is oil- or water-wet.

Figure 5 shows differently shaped relative permeability curves for the two phases. The oil relative permeability curve is concave upward while the gas relative permeability curve has an "S" shape. This figure also shows that the oil relative permeability at the irreducible

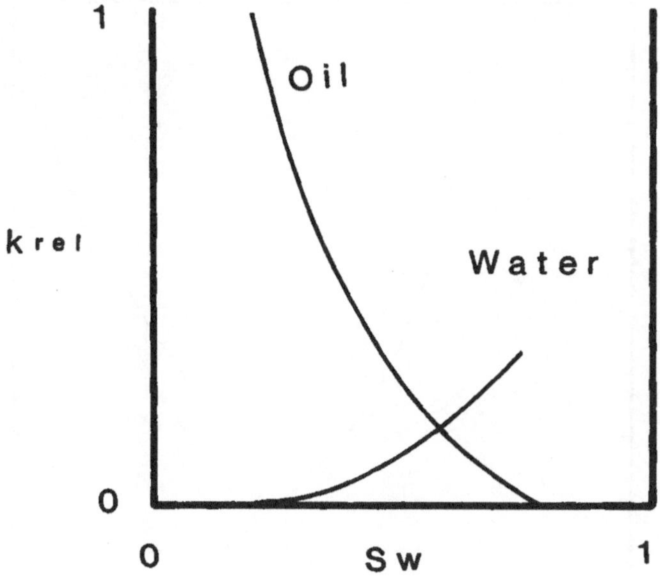

FIGURE 3. Normalized water-oil relative permeability curves.

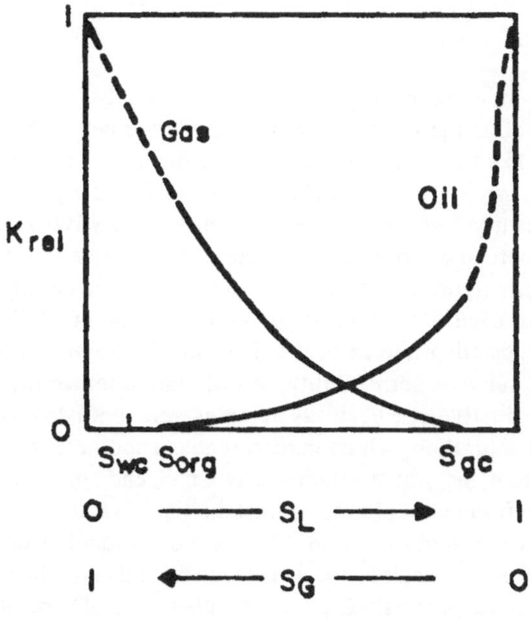

FIGURE 4. Gas-oil relative permeability curves.

(or critical) gas saturation is less than the gas relative permeability at the irreducible oil saturation. Leverett's work[4] shows that the same general observations apply to water-oil relative permeability data. That is, in the presence of oil, the water relative permeability curve takes on the shape of the wetting-phase relative permeability curve or is concave upward.

The shape of the oil relative permeability curve in Figure 5 indicates that, for a small reduction in oil saturation, there is a sizeable decrease in relative permeability to oil. This rapid decline is due to the occupation of larger pores or flow paths by the gas phase. Figure

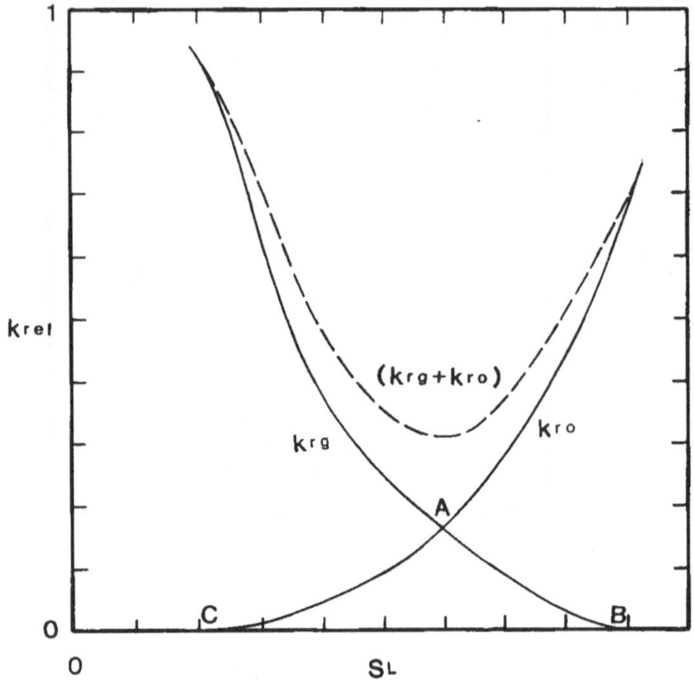

FIGURE 5. Relative permeability curves for an unconsolidated sand.[3]

5 also indicates a steep increase in the gas relative permeability as the gas saturation increases above point "A", which is the saturation at which relative permeabilities to the oil and gas phases become equal. For this unconsolidated sand, the oil relative permeability at 59% oil saturation is equal to gas relative permeability at 41% gas saturation. The gas relative permeability reaches nearly 100% at a gas saturation less than 100%, which means that part of the interconnected pore space does not significantly contribute to the gas permeability of the porous medium. This figure also shows that the gas relative permeability remains at zero until the gas saturation reaches the critical gas saturation, point "B". The gas phase is not mobile at a saturation less than the critical value, but this immobile gas impedes the flow of oil and reduces oil relative permeability. As oil saturation is increased from an initial value of zero, the oil relative permeability remains zero until the oil forms a continuous phase at the critical oil saturation, which is represented as point C in Figure 5. In a solution-gas-drive reservoir, often the water saturation is small and immobile. Therefore, relative permeability values are frequently plotted against the liquid saturation rather than the wetting saturation. Under such a condition, point "C" is the summation of the irreducible water saturation and the residual oil saturation, as previously indicated in Figure 4.

The sum of the relative permeabilities for all phases is almost always less than unity because of interference among phases sharing flow channels. There are a number of reasons for this interference. One of these reasons is that part of the pore channels available for flow of a fluid may be reduced in size by the other fluids present in the rock. Another reason is that immobilized droplets of one fluid may completely plug some constrictions in a pore channel through which another fluid would otherwise flow. Also, some pore channels may become effectively plugged by adverse capillary forces if the pressure gradient is too low to push an interface through a constriction. A fourth reason is the trapping of a group of globules that are clustered together and cannot be moved, since the grain configuration allows fluid to flow around the trapped globules without developing a pressure gradient sufficient to move them. This is the phenomenon that has been referred to as the Jamin effect.

Nowak and Krueger[5] tested two cores in which the permeability to oil in the presence of interstitial water was considerably greater than single-phase permeability to synthetic formation water. Yuster[6] and Odeh[7] both found the same phenomenon based on the results of other work. A possible explanation for the high permeability to oil is that the distribution of clay varies within the rock and variations in water saturation cause variations in the area of contact between water and clay minerals. Thus, increasing degrees of clay swelling may occur at higher water saturation due to the hydration of larger amounts of clay minerals.

Relative permeability is dependent upon both the fluid saturation and the distribution of the various fluids in the interstices of the porous network. This distribution is directly related to wettability characteristics of the rock, which in turn give rise to capillary pressure phenomena. It is well known that hysteresis exists in capillary pressure-saturation curves; therefore, hysteresis in relative permeability-saturation curves can also be expected. Thus, for a given wetting-phase saturation, the relative permeability measured in a rock that is imbibing the wetting phase is not the same as that measured while the rock is draining. Relative permeability values also may be functions of factors such as temperature, overburden pressure, phase equilibria,[161] etc.

III. EFFECTS OF SATURATION STATES

Saturation is a term used to describe the relative volume of fluids in a porous medium. At low saturations of the fluid that preferentially tends to wet the grains of a rock, the wetting phase forms doughnut-shaped rings around the grain contact points. These are called pendular rings. The rings do not communicate with each other and pressure cannot be transmitted from one pendular ring to another. Sometimes such a distribution may occupy an appreciable fraction of the pore space. The amount depends upon the nature and shape of individual grains, distribution, as well as degree and type of cementation.

Above the critical wetting-phase saturation, the wetting phase is mobile through a tortuous path under a pressure differential and as the wetting-phase saturation increases, the wetting-phase relative permeability increases as well. The wetting-phase saturation distribution in this region is called funicular and up to a point, the relative permeability to the wetting phase is less than the relative permeability to the nonwetting phase due to the adhesion force between the solid surface and wetting fluid, and the greater tortuosity of the flow path for the wetting phase. The nonwetting phase moves through the larger pores within this range of saturation, but as the saturation of the wetting phase further increases, the nonwetting phase breaks down and forms a discontinuous phase at the critical nonwetting phase saturation. This is called an insular state of nonwetting-phase saturation.

Fluid flow studies have shown that when immiscible fluids flow simultaneously through a porous medium, each fluid follows its own flow path. This flow network changes for different ranges of saturation and as the nonwetting phase saturation reduces, the network for this phase breaks down and becomes discontinuous; the remaining stationary islands of the nonwetting phase cannot be displaced at pressure gradients encountered in hydrocarbon reservoirs. This condition is referred to as a residual nonwetting phase saturation. Similarly, as the wetting phase saturation decreases, the network through which this phase flows breaks down and becomes discontinuous and immobile. This is referred to as an irreducible wetting-phase saturation.

It has been shown[8-11] that for strongly water-wet unconsolidated sands the permeability to the wetting phase is dependent solely upon its own saturation, (i.e., a plot of k_{rw} as a function of S_w has the same shape regardless of whether or not the pore space contains gas as well as oil). However, in the petroleum related literature, some small[12,13] and some quite large deviations are seen from these findings for consolidated rocks. Some publications[14,15] indicate that the nonwetting phase relative permeability depends on the wetting as well as

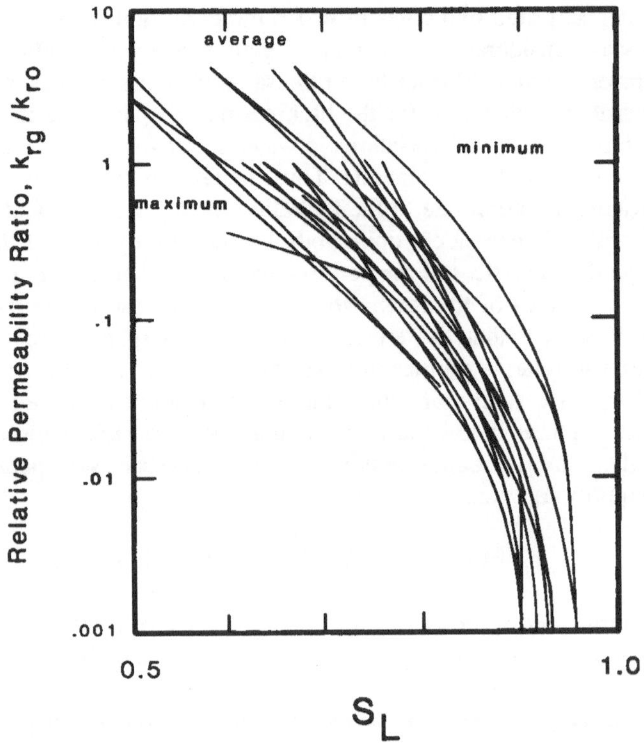

FIGURE 6. Relative permeability ratios for sands and sandstones.[18]

the nonwetting phase saturation for strongly water-wet systems. In preferentially oil-wet systems, the oil phase relative permeability is found to be strictly a function of oil saturation,[16] while in water-wet rocks, the oil phase relative permeability is found to depend on both water and oil saturation. Donaldson and Dean[17] have pointed out that under two-phase flow, relative permeability to water was increased when oil, rather than gas was the nonaqueous phase, indicating that water relative permeability is not solely a function of water saturation.

IV. EFFECTS OF ROCK PROPERTIES

Relative permeability-saturation relations are not identical for all reservoir rocks, but may vary from formation to formation and from one portion to another of a heterogeneous formation.

Arps and Roberts[18] have presented plots of gas-oil relative permeability ratios for 16 consolidated sandstones and 25 dolomites, cherts, and limestones, all with 15% connate water saturation. These plots are presented as Figures 6 and 7. The maximum curve in Figure 6 seems to be typical of unconsolidated sandstone, while the minimum curve appears to be more representative of highly cemented sandstones. The average curve can be considered typical of the average consolidated sandstone. The minimum curve in Figure 7, which seems to be the steepest and most unfavorable, is from a fractured chert core; at the other end of the range, no well-defined maximum case is apparent. Curve #23, adapted from Bulnes and Fitting's work[19] representing 26 samples of west Texas Permian dolomite, appears to be the best maximum curve. The curve selected as ''average'' on Figure 7 appears to be typical of vugular limestones.

Bulnes and Fitting as well as Stone[20] have shown that the fluid flow behavior in uniform-porosity carbonate samples is similar to fluid flow behavior in consolidated sandstones, but the difference becomes pronounced as the rock heterogeneity increases.

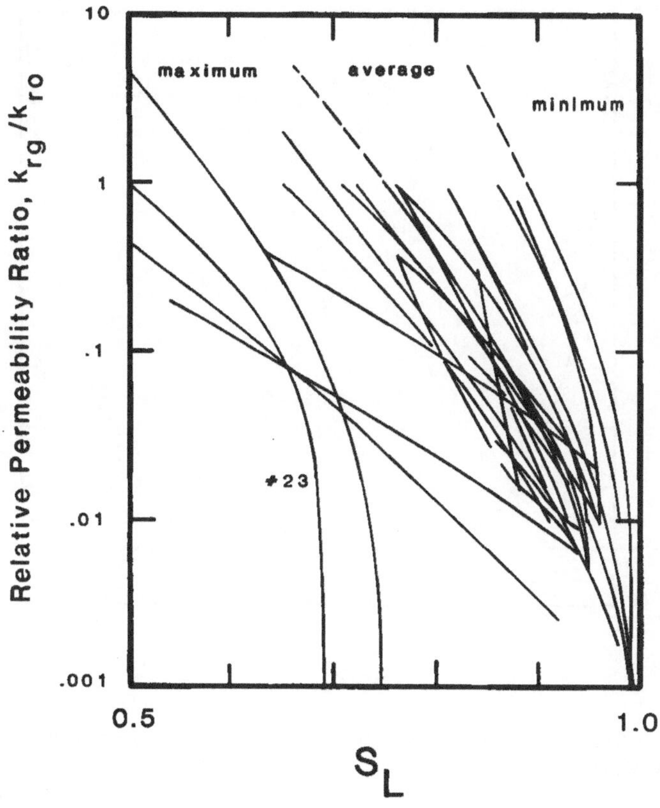

FIGURE 7. Relative permeability ratios for limestones, dolomites, and cherts.[18]

Various works[8,19,21] have shown that the gas-oil relative permeability of consolidated sandstone is qualitatively similar to the gas-oil relative permeability of unconsolidated sand and there is a very close correspondence of the two relative permeabilities to oil at high oil saturation. It has been found that for consolidated sand, the wetting-phase relative permeability drops sharply and the nonwetting phase relative permeability rises steeply as the wetting-phase saturation decreases. However, Naar et al.[22] have shown that there are both qualitative and quantitative differences between relative permeability of consolidated and unconsolidated sands. Owens and Archer[11] indicated that packing as modified by cementation and consolidation affects the equilibrium saturation to the wetting phase but has a negligible effect on the equilibrium saturation of the nonwetting phase. Nind[23] stated that an increase in degree of consolidation increases the nonwetting phase relative permeability in a gas-oil system. Several investigators have noted that the saturation range for a mobile fluid phase is wider in unconsolidated rock than in consolidated rock.

Corey and Rathjens[24] studied the effect of rock heterogeneity on drainage gas-oil relative permeability. They investigated the flow parallel and perpendicular to obvious stratification in anisotropic Berea sandstone cores and concluded that the relative permeability at a given saturation for flow parallel to bedding was greater than the analogous value for flow perpendicular to the bedding plane, as shown in Figures 8 and 9. Huppler[25] found that the water-oil relative permeability of composite core changes appreciably when the sections are arranged in different orders. Johnson and Sweeney[26] also studied the effect of rock heterogeneity on the gas-oil relative permeability ratio.

Leverett[4] found a small but systematic change in the position of the relative permeability-

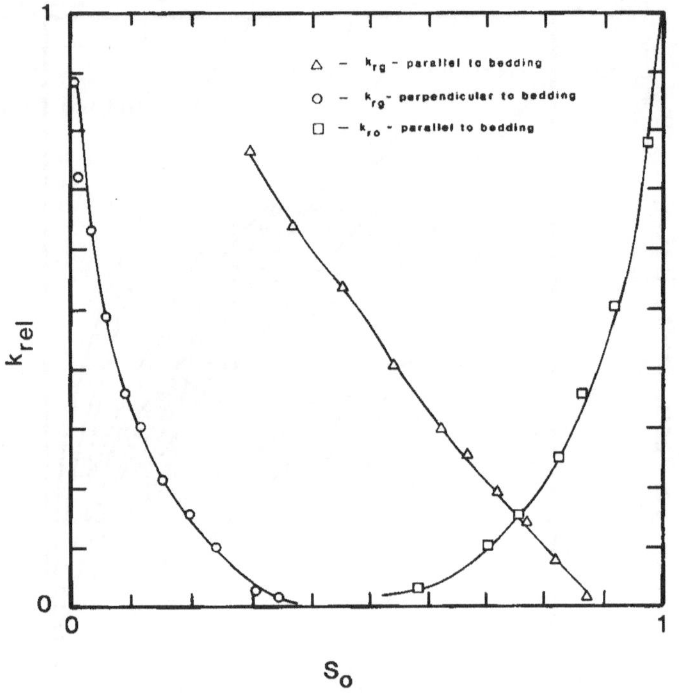

FIGURE 8. Relative permeability measurements from an anisotropic sandstone.[24]

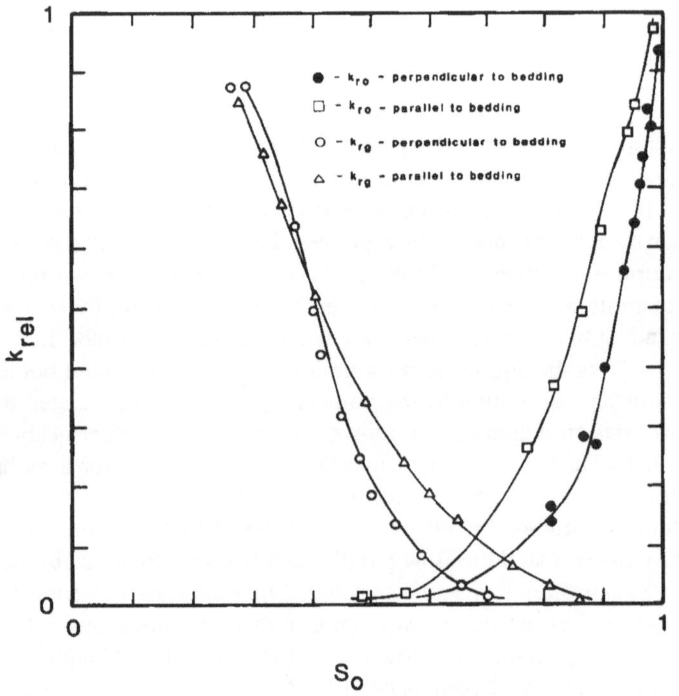

FIGURE 9. Relative permeability measurements from a Berea sandstone.[24]

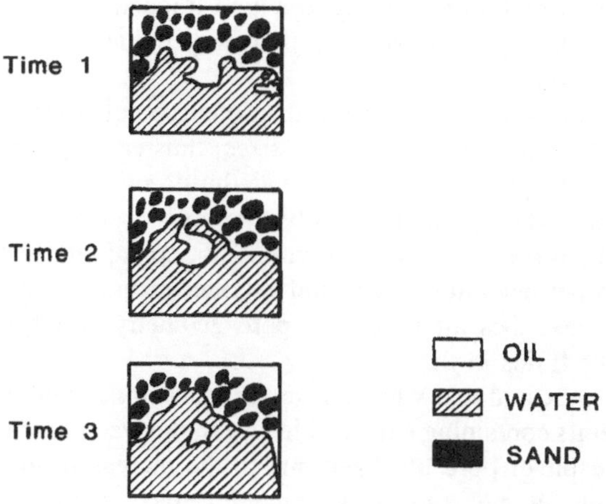

FIGURE 10. The formation of residual oil by the blocking process.

saturation relationship due to the employment of different sizes of sand grains in his experiments. Botset[21] confirmed Leverett's finding and concluded that the effect of grain size distribution was not negligible either on the relationship between relative permeability and saturation or on the value of the equilibrium saturation. It was found that the shape[11] (sphericity), roundness[11] (angularity), and orientation[24] of the grains tended to influence both the shape of the relative permeability curve and the critical gas saturation value in gas-oil systems.

Leverett[4] pointed out that the relative permeability of an unconsolidated sand to an oil-water mixture is related to the sand pore size distribution. Muskat et al.[27] suggested that it is necessary to know the pore geometry of a reservoir rock before fluid movement through it can be analyzed. Morgan and Gordon[28] found that pore geometry and surface area per unit volume influenced water-oil relative permeability curves. They have shown that rocks with large pores and correspondingly small surface areas have low irreducible water saturations and therefore have a relatively large amount of pore space available for the flow of fluids. This condition allows high relative permeability end points to exist and allows a large saturation change to occur during two-phase flow. Correspondingly, rocks with small pores have larger surface areas per unit volume and they have irreducible water saturations that leave little room for the flow of hydrocarbons. This condition creates a low initial oil relative permeability as well as a limited saturation range for two-phase flow.

Gorring[29] demonstrated that oil in a larger pore can be surrounded and blocked off when it is encircled by smaller pores which imbibe the displacing water by capillary forces. He concluded that both pore size distribution and pore orientation have a direct effect on nonwetting residual equilibrium saturation, as shown by Figure 10; therefore, a perfectly uniform packing of spheres should give a residual saturation near zero. Gorring also identified the size of channels occupied by the nonwetting phase as an important factor influencing relative permeability. Crowell et al.[30] indicated that higher initial water saturation yields a higher probability for the nonwetting phase to be in larger channels so that it can be recovered efficiently during wetting-phase imbibition.

Botset[21] mentioned as early as 1939 that the relative permeability-saturation relation depends on the degree and the type of interconnections of the pores. Fatt,[31] Dodd and Kiel,[32] and Wyllie[33] also concluded that the relative permeability of porous media is a direct consequence of the network structure of the media. Pathak et al.[34] concluded that the ratio of pore size to pore throat is a factor which controls the snapping-off of droplets of the nonwetting phase, with a high ratio leading to a high trapped oil saturation. Other workers have investigated the possibility of describing porous media as a network of interconnected pore bodies and pore throats.

Postdepositional alterations can form more than one type of reservoir rock from a single original rock type. Alteration may reduce pore sizes, thus causing higher irreducible water saturation and a narrow range of saturation change during two-phase flow. The presence of grains such as feldspar, when partially dissolved, improves the reservoir rock quality by forming pores larger than the pores between grains not containing feldspar. This alteration causes higher relative permeability values and a larger saturation range during two-phase flow.[28] Reference 35 describes alterations in pore geometry which can occur due to the introduction of reactive fluids in the rock.

Land and Baptist[36] indicated that when a reservoir sandstone contains montmorillonite or mixed-layer clay minerals containing expandable layers, the water sensitivity of the sandstone is not necessarily a result of pore blockage due to the increased volume occupied by the swollen montmorillonite. Some sandstones containing trace amounts of clay minerals may exhibit sensitivity to water resulting from dispersion and subsequent transportation of clay minerals to pore constrictions. Thus, permeability reduction may occur in formations that do not contain expandable clay minerals; however, all formations containing expandable clays are probably water-sensitive due to the ease of dispersion and expansion of this type of clay. Permeability reduction in sands containing sodium clays is likely to be higher than the reduction in sands containing calcium clays.

Some rock properties that influence relative permeability variations are readily observable with a binocular microscope or even more clearly under a scanning electron microscope. Therefore, microscopic core examination can be highly useful for evaluating relative permeability characteristics. Once the significant rock property variations have been identified, a reservoir can be subdivided into appropriate reservoir rock types. Within each of such reservoir rocks types, relative permeability characteristics are usually similar, varying only slightly for rather large changes in air permeability or median grain size.

V. DEFINITION AND CAUSES OF WETTABILITY

"Wettability" is a term used to describe the relative attraction of one fluid for a solid in the presence of other immiscible fluids. It is the main factor responsible for the microscopic fluid distribution in porous media and it determines to a great extent the amount of residual oil saturation and the ability of a particular phase to flow. The relative affinity of a rock to a hydrocarbon in the presence of water is often described as "water-wet", "intermediate", or "oil-wet". Examples of formations with strongly water-wet, strongly oil-wet, and intermediate wettability are the Spraberry formation in west Texas, the Black Bradford sand in Pennsylvania, and the Fairbank sand in south Texas, respectively.

Wettability may be represented by the contact angle formed among fluids and a flat solid surface or the angle formed between the fluids' interface and a glass capillary tube, as shown by Figure 11. The angle is measured through the denser fluid.

The wettability of a porous medium is determined by a combination of all surface forces. A sketch is shown in Figure 12, wherein two liquids, oil and water, are in contact with a solid. The force exerted by water to spread laterally and displace oil (interfacial tension between water and oil) is opposed by the resultant of the solid and liquid forces (solid-oil and solid-water interfacial tensions). This difference in opposing forces is called the adhesion tension:

$$A_t = \sigma_{so} - \sigma_{sw} = \sigma_{wo} \cos \theta_{wo} \tag{2}$$

This relationship is referred to as the Young-Dupre equation, where A_t is the adhesion

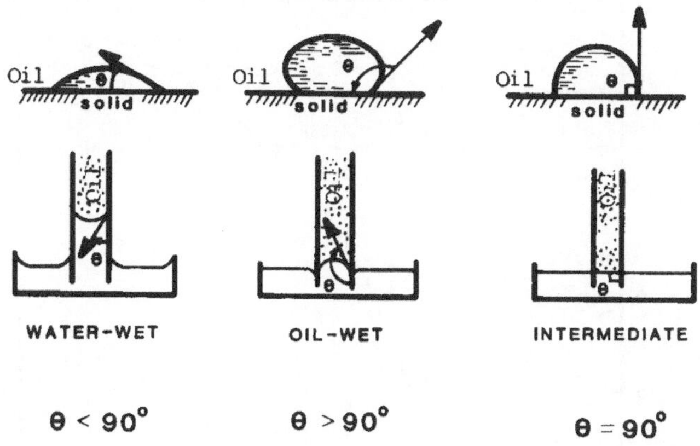

FIGURE 11. Wettability conditions on flat surfaces and in capillary tubes.

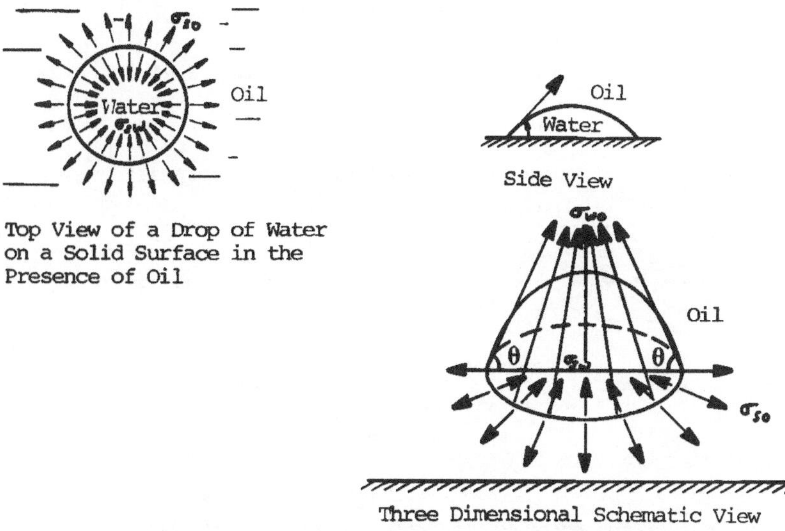

FIGURE 12. Forces at a water-oil-solid interface.

tension; σ_{so}, σ_{sw}, and σ_{wo}, respectively, are solid-oil, solid-water, and water-oil interfacial tensions (usually measured in dyne/cm); θ_{wo} is the contact angle between water and oil measured through the denser liquid phase (usually water).

A positive value of adhesion tension means the contact angle is less than 90° and the solid surface is preferentially water-wet. A zero value of adhesion tension indicates that the contact angle is equal to 90°; this is intermediate wettability.

A negative value of adhesion tension means the contact angle is greater than 90° and that the solid surface is preferentially oil wet. There is no practical laboratory method for measuring σ_{so} or σ_{sw}. However, σ_{wo} and cos θ are measurable quantities which can be used to evaluate the wettability of a solid surface. A fluid is referred to as wetting or nonwetting to a surface depending on whether the contact angle is less than or greater than 90°.

Understanding the causes of wettability requires a study of the chemistry of the fluids, the polarity and molecular weight of reservoir hydrocarbon compounds, and the occurrence of surface chemical processes at the solid-fluid interfaces.

Stegemeier and Jensen[37] experimentally found that the contact angles vary directly with molecular weight for liquids with similar chemical structures. Figure 13 shows this variation for the normal paraffin series compounds.

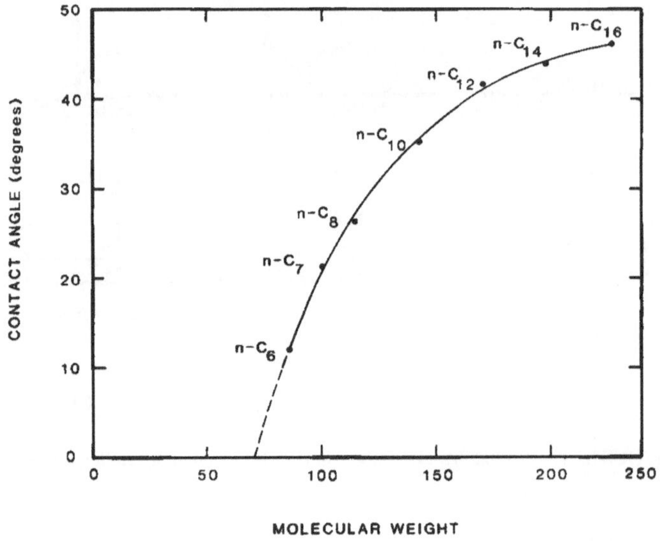

FIGURE 13. Contact angle as a function of molecular weight.[37]

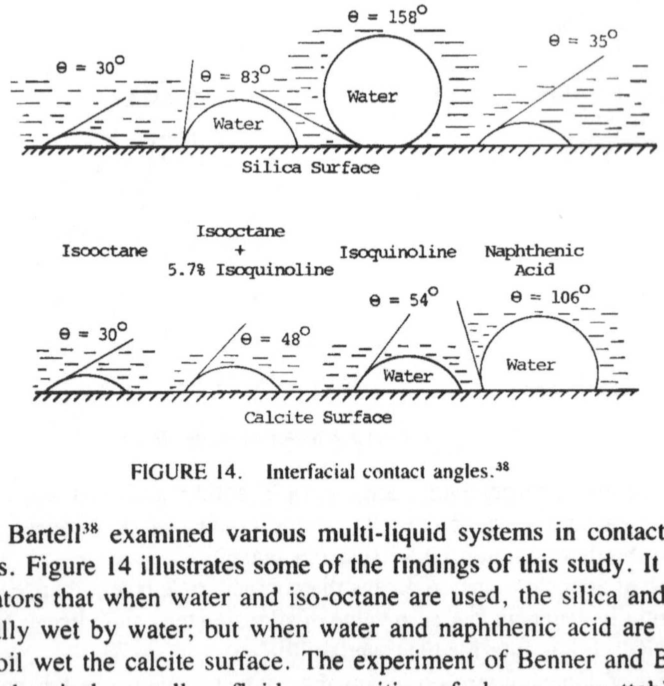

FIGURE 14. Interfacial contact angles.[38]

Benner and Bartell[38] examined various multi-liquid systems in contact with silica and calcite surfaces. Figure 14 illustrates some of the findings of this study. It was reported by these investigators that when water and iso-octane are used, the silica and calcite surfaces are preferentially wet by water; but when water and naphthenic acid are used, water wet the silica but oil wet the calcite surface. The experiment of Benner and Bartell illustrated the effects of chemical as well as fluid composition of phases on wettability of a porous medium. Contact angles as low as 30° and as high as 158° were observed when various chemicals were employed in the study.

Salathiel[39] discovered that the wettability of mineral surfaces may be altered not only by adsorbed monolayers of surface-active polar compounds, but also by much thicker layers of deposited organic materials. Several other workers have reported the formation of stable films on solid surfaces when the surfaces stand in contact with certain crude oils. Reisberg and Dosher[40] described the deposition on glass or quartz surfaces of highly stable and appreciably thick films of strongly oil-wet material from Ventura crude oil.

Early experimenters thought that all oil-bearing formations were strongly water-wet because an aqueous phase was always the fluid initially in contact with reservoir rock; furthermore, silica and carbonates are normally water-wet in their clean state. Subsequent studies suggested that many oil reservoirs are not strongly water-wet and that the presence of crude oils containing natural surface-active agents, such as asphaltic or wax type material readily adsorbable by solid-liquid interfaces, can render the solid surface oil-wet.[41] Other studies provide evidence that reservoir rock wetting preference may cover a broad spectrum.

One criticism of the idea of reservoir rock surfaces becoming modified by the adsorption or deposition of polar organic material from the oil phase is that such materials should have been eliminated during migration from the source rock to the reservoir. On the other hand, geochemists are now finding substantial evidence of various alteration processes which affect crude oils subsequent to their accumulation in reservoirs. In a discussion of natural gas deasphalting, Evans et al.[42] suggested a reasonable hypothesis that the more gas a crude has in solution the more of its heavy ends have come out of solution, plating out on the reservoir rock. It may be noted in this respect that Salathiel's strongly oil-wet film deposition on quartz and porous rocks from a mixture of evacuated crude oil and heptane was also probably the result of a deasphalting process.

Despite uncertainty as to the causes of reservoir wettability, much evidence has been presented in recent years to suggest that many oil reservoirs are not strongly water-wet. In particular, there are the many brine/crude oil contact-angle measurements of Treiber et al.[62] and the conclusions of Salathiel with regard to the apparent wetting characteristics of the Woodbine reservoir in the East Texas Field. Nutting[43] as early as 1934 indicated that some reservoir rocks are oil-wet. Leach et al.[44] described a reservoir believed to be oil-wet. Mungan[45] studied fresh carefully preserved cores from a reservoir and concluded that the formation was oil-wet. Schmid[46] has shown that strongly water-wet cores became less water-wet when equilibrated with some crudes. Kusakov et al.[47] studied the thickness of a water film left on a quartz surface under crude oil drops and found that for two of the crudes, the film will rupture, bringing the crude oil into direct contact with the quartz surface; the surface can then be described as water-wet at some spots and oil-wet at others. Also, Craig[48] suggested that most formations are of intermediate wettability with no strong preference for either oil or water. There is recent evidence to suggest that water may not always completely wet reservoir rock in gas-water flow following solvent injection. Soil scientists concerned with air/water/soil systems have reported situations in which there is incomplete wetting by the aqueous phase.[49]

Authors such as Holbrook and Bernard,[50] and Fatt and Klikoff[51] assumed that wetting of reservoir solids was heterogeneous rather than uniform. Holbrook and Bernard measured fractional wettability by dye adsorption. Brown and Fatt[52] defined fractional wettability as the fraction of surface area in contact with water. This may not be a constant value since the water and oil saturations change as a reservoir is produced. Schmid[46] showed by means of capillary pressure-saturation data, that in preserved cores the fine pores were water-wet while the large pores were much less water-wet. This type of wetting is often referred to as "spotted", "dalmation", or "fractional". That heterogeneous wettability is a normal condition in oil sands has also been suggested by Salathiel,[39] Iwankow,[53] Brown and Fatt,[52] Gimaludinov,[54] and McGhee and Crocker.[55] Several of these investigators have suggested that the wetting phase completely occupies the smaller pores of a reservoir rock in addition to the rock surface of the larger pores, while the nonwetting phase primarily occupies the insular regions of the larger pores. Evidence suggests that some oil reservoirs are partly preferentially water-wet and partly preferentially oil-wet. Such a condition could arise if some pores are lined with one type of mineral and other pores are lined with another mineral.

The existence of different minerals in porous media can create differences in surface chemistry of the grains, so all grain surfaces do not have the same affinity for surface active compounds. For instance, a tertiary sand reservoir in Alaska contains quartz and siderite minerals which are strongly water-wet and calcite which is strongly oil-wet. The overall

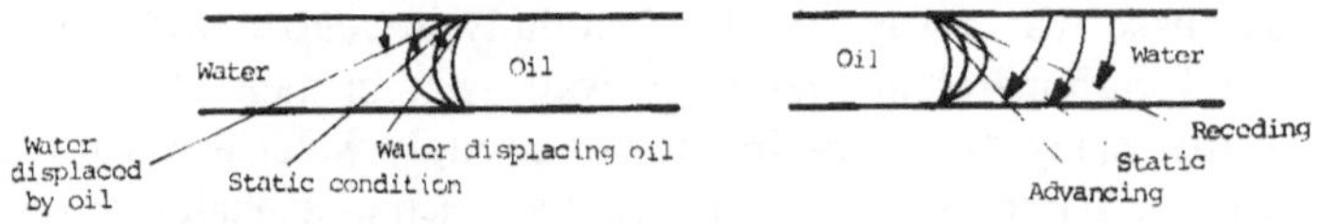

FIGURE 15. Advancing and receding contact angles in capillary tubes.

rock system is water-wet, probably due to the presence of quartz and siderite surfaces in the main flow channels. The presence of anhydrite or gypsum in the flow channels of some carbonate rock may alter its wettability. These minerals are found to create a strongly water-wet system, while many carbonate rocks are probably oil-wet under reservoir conditions. Heavy metal sulfides are known to render a surface oil-wet when they are present in the flow channels of porous media.

Wagner and Leach[56] stated that in some oil reservoirs the rock surface is covered by a firmly attached bituminous or other organic coating. Such surfaces would be preferentially oil-wet in the presence of oil and water, regardless of oil and water composition. Boneau and Clampitt[57] reported that the oil-wet character of the North Burbank reservoir is due to a coating of chamosite clay which covers approximately 7% of the quartz surface.

VI. DETERMINATION OF WETTABILITY

The wettability of a rock can be either evaluated experimentally or estimated qualitatively. There is no satisfactory method to determine *in situ* reservoir wettability. However, laboratory-measured wettability has been used to evaluate *in situ* wettability. Many of the widely used experimental methods of wettability evaluation utilize either the reservoir rock or the reservoir fluids, but not both. Therefore, a laboratory wettability evaluation should be related to actual reservoir conditions using a great deal of caution.

A. Contact Angle Method

The contact angle method is used by a number of laboratories; the technique has received considerable attention in the literature as a quantitative method of wettability measurement. The method consists of measuring the contact angle θ that a drop of pure liquid resting on a smooth, flat, incompressible, nonporous, homogeneous solid forms when immersed in another fluid. In most practical situations, the contact angle formed between the solid surface and the water-oil interface is found to exhibit two limiting values rather than a single equilibrium value. The value of the contact angle when water is brought into contact with oil on a solid surface previously in contact with oil is called the "advancing contact angle". The value of contact angle when oil is brought into contact with water on a solid surface previously in contact with water is called the "receding contact angle".

Figure 15 shows a comparison of advancing and receding contact angles in a capillary tube. The fact that advancing and receding contact angles are not equal is referred to as contact angle hysteresis and it is usually attributed to surface heterogeneity and roughness, as well as the presence of surface-active materials[58] and rate of fluid movement. As the surface roughness of a rock increases, the contact angle will further increase, provided the contact angle measured on the smooth surface of the rock is above 90°; however, if the contact angle measured on a smooth surface is less than 90°, the increase in surface roughness would further decrease the angle. The smooth surface contact angle is found to increase in advancing and decrease in receding, on the rough surface over most of the 0 to 180° contact angle range.[59]

Surface-active materials in the fluids may cause adsorption processes at the solid-fluid interfaces which give rise to appreciable contact angle hysteresis even with a smooth, homogeneous solid. Motion of the three-phase line of contact increases contact angle hysteresis as the rate of movement increases.

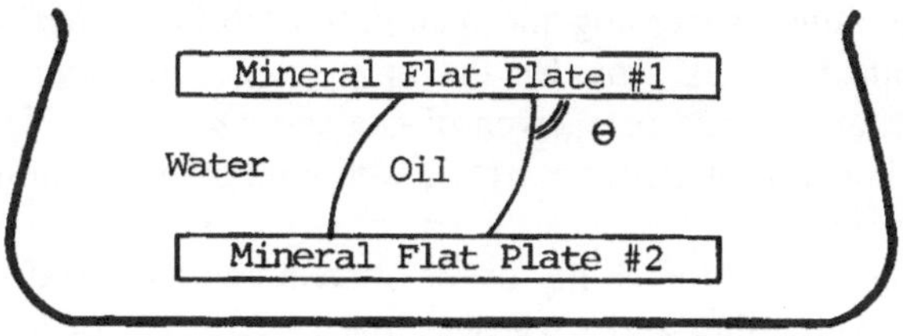

FIGURE 16. Schematic measurement of contact angles.[56]

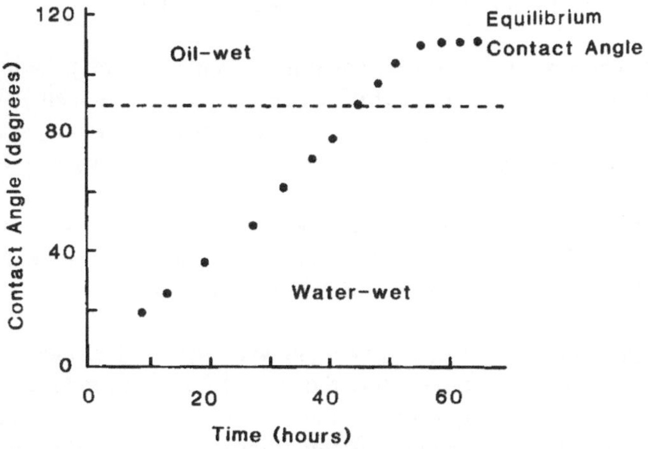

FIGURE 17. Influence of aging on laboratory-measured contact angle.[48]

Advancing and receding contact angles can be shown in a capillary tube for oil displacing water (receding angle) and water displacing oil (advancing angle). The procedure to determine the contact angle using a contact angle cell is described by Wagner and Leach[56] and is illustrated schematically by Figure 16. Briefly, samples of polished, flat plates of the mineral which is the main constituent of the reservoir rock are immersed in a sample of formation water. A drop of reservoir oil is held between the two flat samples of the mineral and the two plates are moved horizontally so that the water advances on the surface of the plate initially covered by oil. The contact angle formed between the interface and the newly water-occupied surface of the mineral is a measure of the water advancing contact angle. The advancing contact angle is the one that is customarily measured and often reported without being identified as advancing.

The contact angle measured in the laboraotry is often influenced by aging. It has been shown that contact angle increases with age of the oil-solid interface until an equilibrium is reached. This may require several days and it is one of the disadvantages of the contact angle method.[48] Figure 17 shows this effect.

Reliable wettability measurement requires that both the reservoir rock and the fluids be free from contaminants. Uncontaminated reservoir rocks can probably be obtained if the cores are recovered with coring fluid containing no surface-active additives or with reservoir oil that has not been exposed to oxygen. It has been reported that exposure of cores to air could result in alteration from water-wet to intermediate wettability. Uncontaminated reservoir water and oil are easier to obtain than unaltered reservoir rock. Since contact angle measurement can be done without a sample of (uncontaminated) reservoir rock, it has become a widely used method for determining wettability.

Zisman[60] and other investigators studied contact angles under controlled conditions and expressed varying opinions concerning the method's usefulness. Melrose and Brandner[61] believed that the contact angles provides the only direct and clear specification of the wettability property characteristic of a given oil-water-rock system. Treiber et al.[62] found that the water-advancing contact angles correlate well with other wettability indicators while water-receding angles do not.

Brown and Fatt[52] questioned the ability of the contact angle method to provide a reliable scale for determining wettability and suggested that the concept of a contact angle representation of wettability of reservoir rock be abandoned and that this method be replaced with a "fractional surface area" method. Morrow et al.[63] also observed that several factors cast doubt on the utility of the contact angle method. Mungan[64] described some of the limitations and pitfalls of contact angle measurement as follows:

1. The mineral chosen for the contact angle measurement is the principal constituent of the reservoir rock. For the purpose of contact angle measurement, silica or quartz is used to represent a sandstone; calcite is used to represent a carbonate or reef reservoir. Laboratory measurement of contact angle or mineral surfaces may not simulate true reservoir contact angle.
2. The contact angle at the water/oil displacement front is "advancing" while at the leading edge of the oil bank it is "receding". These values sometimes differ by as much as 50°. This variation can be on the same order of magnitude as the laboratory-measured contact angle.
3. Contact angle measurement should be done when the solid surface and a fluid remain in contact for an adequate time before the second fluid is introduced over the surface. This is referred to as pre-equilibrium time and it is of different length for each crude oil-water system. Without adequate pre-equilibrium time, a stable contact angle is not reached. In some cases it has been reported that a stable contact angle is never obtained if the solid surface comes into contact with some types of crude oils. Contact angle measurement is frequently time consuming.
4. Contact angle measurement should be performed with actual reservoir fluids, since they are in equilibrium and solubility effects are negligible; otherwise, the fluids must be equilibrated with one another so that the solubility effects become negligible.
5. Contact angle measurement preferably should be done with bottom-hole fluid samples; however, because of the time and expenses involved, flow line samples are often used. Fluid samples taken from the storage or treating facilities are not reliable, due to the possible accumulation of asphaltenes. When produced water is not available, synthetic brine is commonly used.
6. Contact angle measurements should be made under controlled conditions so that the oxidation of crude oil can be prevented.
7. Contact angle measurement requires extreme care to assure cleanliness and inertness of the apparatus.

B. Imbibition Method

An imbibition test is a reliable technique of wettability determination provided unaltered reservoir fluids are available. The method consists of the measurement of rate of flow of a wetting fluid spontaneously imbibed into a core and replacing a nonwetting fluid by the action of capillary forces alone.

Imbibition tests may be performed at standard conditions or at reservoir conditions. Figures 18, 19, and 20 illustrate equipment that is used for conducting the tests at ambient conditions. The imbibition test at standard conditions may be performed as follows:

1. A cylindrical plug of reservoir rock 1 to 1 $^1/_2$ in. in diameter is cut with water as a coolant in the cutting process.

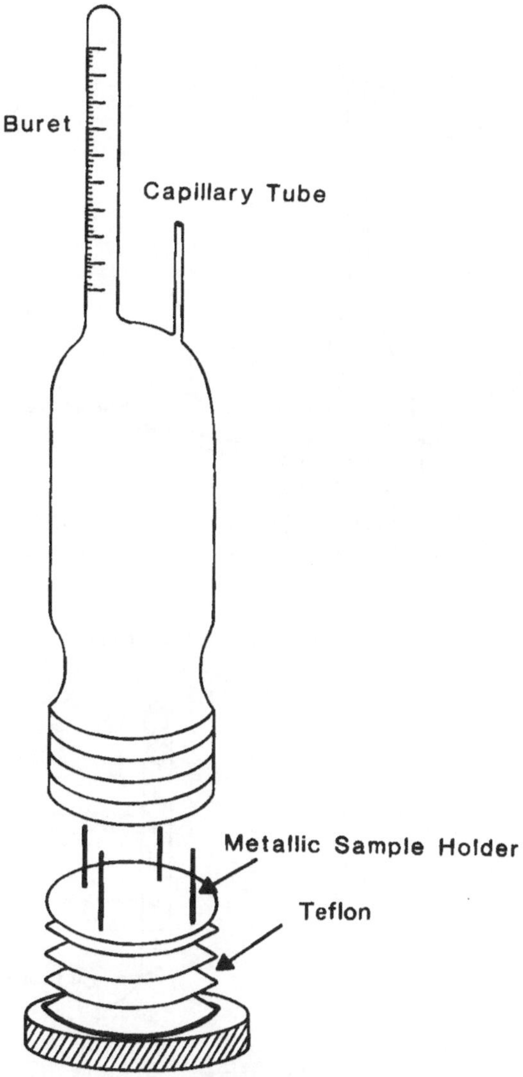

FIGURE 18. Imbibition cell.

2. The sample is placed under water in a beaker and evacuated to remove trapped gas.
3. The sample is flushed with water to reduce the oil saturation to residual level.
4. The core plug is placed in an imbibition cell under oil and oil imbibition is monitored.
5. The drained water is measured; it is equal to the amount of imbibed oil. Sufficient time should be allowed for the system to reach equilibrium; this may take several days depending on the permeability of the plug.
6. The plug is then saturated with oil to reduce the remaining water to the irreducible level.
7. The sample is placed in an imbibition cell under water and water imbibition is monitored by the amount of oil being drained. The fluid that imbibes into the sample (oil or water) is the wetting phase.

The imbibition test under reservoir conditions is more complex. Irreducible water saturation is established by flushing the core with live oil and the imbibition tests are made at reservoir pressure and temperature.

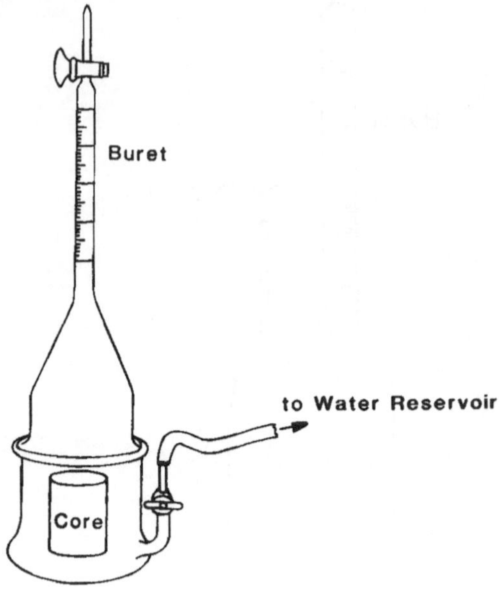

FIGURE 19. Imbibition cell.

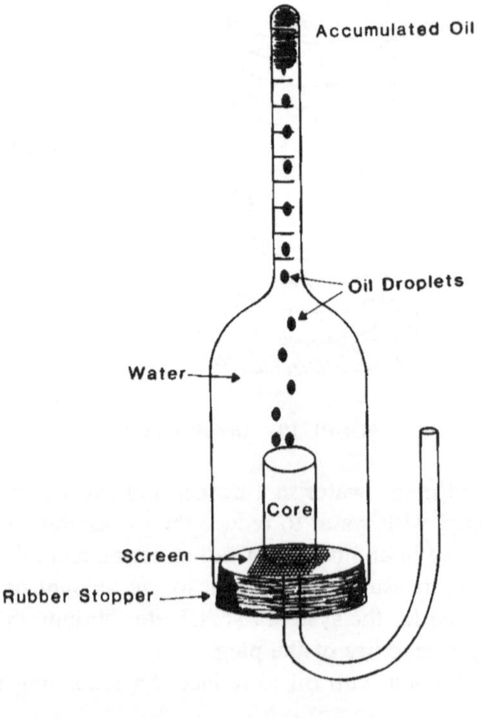

FIGURE 20. Imbibition cell.

Amott[65] developed a quantitative technique for defining the degree of water-wetness of cores. He expressed the degree of water wetness by a water index, which he defined as the ratio of the volume of water spontaneously imbibed into a core to the total volume of oil displaced by a water drive (forced displacement of oil by water). Similarly, an oil index was defined at the ratio of the volume of oil spontaneously imbibed to total water displaced by an oil drive (forced displacement of water by oil). Amott's test consists of the following steps:

1. Flush the reservoir sample with water to reduce the oil saturation to its residual level.
2. Immerse the sample in water and evacuate to remove gas.
3. Immerse the sample in kerosene (or reservoir oil) and measure the volume of water displaced by imbibition of oil after 20 hr.
4. Measure the volume of water displaced when the sample is centrifuged under oil.
5. Immerse the sample in water and measure the volume of oil displaced by water after 20 hr.
6. Measure the volume of oil displaced when the sample is centrifuged under water.

Oil index is the ratio of the volume of fluid measured in step 3 to the volume of fluid measured in step 4. Water index is the ratio of fluid volume from step 5 to fluid volume from step 6.

The preferential wettability of a rock is determined by the magnitude of these two indexes, i.e., strong wettability is indicated by values approaching one and a weak preference in indicated by values approaching zero. A water index of one indicates a strongly water-wet surface while an oil index of one indicates a strongly oil-wet surface. Values between these two extremes or a value near zero for both ratios cover the range of intermediate wettability.

Amott's test of wettability of porous media received high marks from Raza et al.[66], although Moore and Slobad,[67] Bobek et al.,[68] Killens et al.,[69] and Richardson[70] have indicated that the imbibition rate cannot be entirely attributed to the wettability of the core, but that it is also influenced by rock porosity, permeability, pore structure, and pore size distribution, as well as viscosity and interfacial tension of the fluids involved in the experiment. Donaldson et al.[71] tried to eliminate extraneous effects from the wettability measurement by comparing the volumes of fluids imbibed into preserved reservoir cores with the volumes of fluids imbibed in the same cores after extraction and resaturation. Although the use of the same core would appear to offer identical pore size distributions, the change in fluid distributions caused by the cleaning process may have offset the advantage gained.

Mungan[72] reported the use of an imbibition test to evaluate the wettability of native-state cores. Emery et al.[73] used an imbibition test after incubation of cores for up to 1,000 hr with gas-saturated oil under pressure; water was the first phase to contact the rock in the test. Kyte et al.[74] described imbibition tests conducted at reservoir temperature and pressure.

C. Bureau of Mines Method

The U.S. Bureau of Mines method of wettability determination of a porous rock, commonly referred to as the "Centrifuge Method", is based on the assumption that an elemental area of the internal surface of the porous medium is either wettable or nonwettable by one of the fluids involved. The problem is one of determining the fraction of the internal surface wetted by each fluid. A method of measuring wettability based on the above theory was suggested by Gatenby and Marsden[75] and was later developed by Donaldson.[71] These investigators made use of the areas obtained from the drainage and imbibition cycles of the capillary pressure curve to produce a numerical representation of wettability. The Bureau of Mines method is quite rapid and it can be employed with reservoir fluids.

D. Capillarimetric Method

Johansen and Dunning[76] recognized the importance of the liquid used in determining wettability of a rock-liquid-brine system and suggested the use of a capillarimeter which joins the two liquid phases, oil and water, through a small diameter glass capillary tube, with a capillary pressure across the interface joining the two phases. Adhesion tension or displacement energy, was calculated from the difference in height of the two liquids in the two arms of the capillarimeter, the difference in densities, and the acceleration due to gravity. The instrument is capable of measuring interfacial forces with either an advancing or receding interface. Major limitations of this method are the exclusion of reservoir rock as a factor influencing wettability and lack of provision to prevent oil from oxidizing.

E. Fractional Surface Area Method

This method, developed by Brown and Fatt,[52] uses mixtures of untreated sand and sand rendered oil-wet by organosilane vapors to obtain wetting conditions ranging from completely water-wet to completely oil-wet.

Wettability is represented by the fraction of solid surface made artificially oil-wet. Although use of the method to evaluate field behavior is not in evidence, the concept of a fractionally wet surface has been presented in the work of other writers.[39]

F. Dye Adsorption Method

This method, developed by Holbrook and Bernard,[50] is based upon the ability of reservoir rock to adsorb a dye such as methylene blue from aqueous solution, while rock surface areas covered by contaminants from the oil phase remain unaffected. The test is based on a comparison of the adsorption capacity of the test sample with that of an adjacent sample extracted by chloroform and methanol. This method makes assumptions similar to those of Brown and Fatt[52] in their "fractional surface area" method.

G. Drop Test Method

This method is often used to confirm rock wettability. The procedure involves placing drops of oil and water on the surface of a fresh break in the core. The fluid that imbibes is the wetting phase while the fluid that forms a ball and does not wet the surface is nonwetting. The drop test is a qualitative determination and is sometimes misleading.

H. Methods of Bobek et al.

Bobek et al.[68] proposed a laboratory test to ascertain preferential wettability in a qualitative fashion. The technique consists of determining which fluid will displace the other from a rock sample by imbibition. The results of this imbibition test are compared with those of a reference imbibition test on the same core sample after it has been heated to 400°F for 24 hr to remove any organic materials and to make it more water-wet. The assignment of qualitative wettability designations is based on the relative amounts and rates of imbibition in the two tests.

In the same paper a method for estimating the wettability of unconsolidated material is discussed. A thin layer of the unconsolidated sand is spread on a microscope slide. The oil content of the sand is increased by adding a clear refined oil. Droplets of water are then placed on the surface of the sand grains and the fluid movement is observed. If the sand is water-wet, the added water will displace oil from the surfaces of the sand grains and the oil will form spherical droplets, indicating that oil is the nonwetting phase. A similar procedure is used to test for oil wettability.

I. Magnetic Relaxation Method

A nuclear magnetic relaxation technique was suggested[52] for determining the portions of the rock surface area that are preferentially water-wet or oil-wet. A rock sample is first exposed to a strong magnetic field, then to a much weaker field. The magnetic relaxation rate — that is, the rate at which the initially imposed magnetism is lost — is then measured. In sandpacks containing known mixtures of oil-wet and water-wet sand grains, a linear relationship was observed between the relaxation rate and the fraction of the surface area

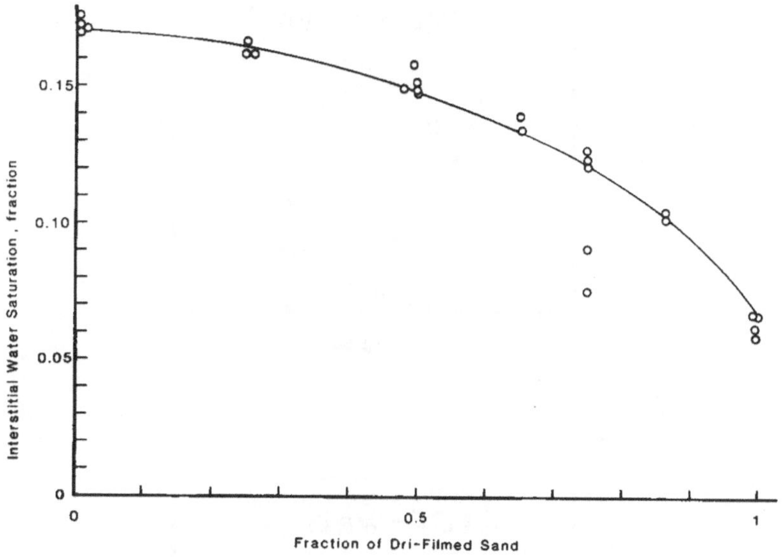

FIGURE 21. Interstitial water saturation for sand mixtures.[53]

that is oil-wet. Though the authors reported no studies using natural cores, they proposed a testing procedure. Their technique requires specialized equipment not normally found in petroleum laboratories and there are no indications in the literature that the method has found routine use.

J. Residual Saturation Methods

McGhee et al.,[55] Lorenz et al.[79] and Reznik et al.[80] reported a correlation between residual oil saturation and wettability. Treiber et al.[62] reported that the connate water saturation in a native core can sometimes be used as an indication of formation wetting preference. They found that oil-wet formation have much lower connate water saturations than the water-wet ones. In addition, the connate water saturation in a strongly oil-wet reservoir was found to be constant regardless of the sample permeability, while in reservoirs of other wettabilities the connate water saturation decreased with increase in permeability. Iwankow[53] also described the effect of heterogenous sand wettability in terms of a fraction of drifilmed sand. (See Figure 21.) Drifilm is a solution commonly used in the laboratory to make sands preferentially oil-wet. Coley et al.[81] were not successsful in using the ratio of the wetting to the nonwetting residual saturation from relative permeability-saturation relationships as a rock preferential wettability indicator; however, they found that the volume of mobile fluid shown by the spread between the residual saturation values of a relative permeability curve appears to decrease as the oil wettability increases.

K. Permeability Method

The determination of wettability of a sample from permeability data is accomplished by comparing the ratio of water permeability at residual oil saturation with the oil permeability at connate water saturation. If this ratio is less than 0.3, the sample is considered to be water-wet, while a value near unity indicates that the sample is oil-wet.[82] The relationship between absolute permeability and connate water saturation has been frequently mentioned in the petroleum literature and the relationship between connate water saturation and rock wettability has been discussed. Rocks with low connate water saturation are considered to be weakly water-wet to oil-wet, while rocks with high connate water saturation are normally designated as water-wet.

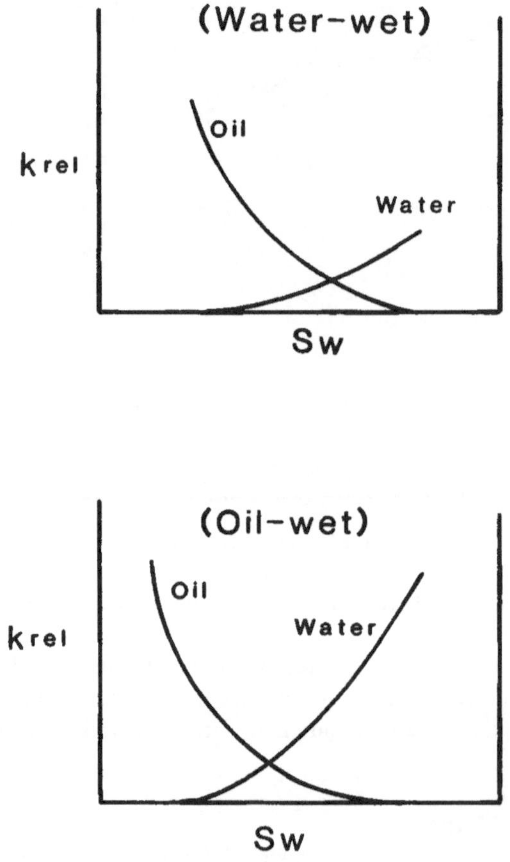

FIGURE 22. Schematic wettability effects on relative permeability curves.

L. Connate Water-Permeability Method

A correlation of absolute permeability as a function of water saturation in cores cut with oil-base mud has been used for qualitative identification of core wettability.[65] Water saturation is measured in freshly cut cores and absolute permeability is determined after extraction and drying. A plot of water saturation as a function of absolute permeability to air is prepared. The curve will have a gentle slope over a large saturation interval for water-wet systems, while it will exhibit a nearly vertical slope over a narrow saturation range for oil-wet systems. This technique is applicable primarily to thick hydrocarbon reservoirs with sufficient variation in permeability and water saturation so the required plot can be prepared.

M. Relative Permeability Method

For a given water saturation, the water relative permeability of a water-wet rock is lower than that of a comparable oil-wet rock. For the systems studied by Owens and Archer[11] it was found that an increase in oil wetness (at constant water saturation) produced an increase in k_{rw} and a decrease in k_{ro}. Treiber et al.[62] concluded that water-wet consolidated porous media normally have a water relative permeability less than 15% at residual oil saturation, while oil-wet porous media show a 50% or higher relative permeability to water at flood-out.

Craig[48] offers the following heuristic guidelines, which are illustrated by Figure 22:

	Water-wet	Oil-wet
S_{wi}	>20 to 25%	<15%, usually 10%
$k_{rw} = k_{row}$	@ S_w>50%	@ S_w<50%
k_{rw} at S_{orw}	<0.3	>0.5, approaching 1.0

In a water-wet rock, residual oil globules in the large flow channels block the easy flow of water and cause a low water relative permeability; however, the oil in an oil-wet system occupies smaller flow channels and coats the walls of the larger ones, causing a minimum disturbance to water flow and a higher water relative permeability.[83] This is why an oil-wet reservoir will waterflood poorly, with early water breakthrough, rapid increase in water cut, and high residual oil saturation.

The water-oil relative permeability relationship of native-state cores under steady-state conditions is one of the best indicators of the rock wettability preference. Keelan[82] pointed out that a sharp drop in oil relative permeability over a small saturation change accompanied by a rapid rise in relative permeability to water, to a terminal value in excess of one third the initial oil relative permeability, often indicates oil wetness. Careful sample examination is essential in using this technique, for heterogeneous or cracked samples yield relative permeability data similar to the data obtained from oil-wet cores.

Water relative permeability curves in water-oil systems show good agreement with the oil relative permeability curve obtained during gas-oil relative permeability tests in a strongly water-wet core.[62,63,84] This effect does not exist under any other wetting condition. In a strongly water-wet core, the water relative permeability curve of a water-oil system also shows good agreement with the water relative permeability of a gas-water system in the presence of residual oil saturation. This agreement will occur, even though the direction of the change in saturation may not be the same in the two systems. In the same manner, in strongly oil-wet cores, the gas relative permeability of a gas-water system is comparable to the gas relative permeability of a gas-water system in the presence of residual oil saturation.[84]

The point of intersection of the water and oil relative permeability curves has been suggested as an indication of rock wettability. Owens and Archer[11] have shown that the relative permeability intersection point moves toward higher values of water saturation and lower values of relative permeability in a water-oil system as the sample wettability is changed from oil-wet to water-wet. As illustrated by Figure 22, a relative permeability intersection point on the left of 50% water saturation indicates that the system is oil-wet, while an intersection to the right of this saturation suggests that the system is water-wet.

N. Relative Permeability Summation Method

The summation of relative permeabilities to the water and oil phase at fixed saturations also gives some insight into the immiscible flow processes. McCaffery[59] noted a trend in the minimum values of the sum of relative permeabilities of samples according to their preferential wettabilities.

O. Relative Permeability Ratio Method

If the ratio of displacing to displaced phase relative permeability is plotted as a function of the displacing-phase saturation, the shape of the plot is related to preferential wettability of the rock.[66] It has been shown that the water-oil relative permeability ratio shifts to a higher value as the rock becomes more oil-wet; furthermore, a semilog plot of water-oil and gas-oil relative permeability indicates that the gas-oil relative permeability ratio curve moves from under to over the water-oil relative permeability ratio curve as the rock becomes preferentially water-wet.[51] The water-oil relative permeability ratio curves of rock with various degrees of intermediate wettability are found to be practically the same in the presence of constant initial water saturation.[85] Imbibition water-oil relative permeability ratio curves in the absence of initial water saturations show higher values of residual oil saturation as the cores become more oil-wet.[85] Steady-state relative permeability measurements should be used for determination of wettability. Unsteady-state methods may not allow equilibrium to occur during the flow test; therefore, they may indicate more oil wettness than actually exists.

P. Waterflood Method

Several attempts to find a single correlation of wettability with waterflood oil recovery for different porous media have failed, even though the tests were carried out under a standard set of conditions.[65] However, the waterflood performance of a native-state core under carefully controlled laboratory conditions has been used as an indication of rock preferential wettability. It is found that in a strongly water-wet system, a large fraction of the oil is produced prior to water breakthrough and very little additional oil is recovered after breakthrough. For the test to be reliable, an equilibrium wetting condition must prevail prior to the passage of the flood front through the core.

Q. Capillary Pressure Method

Both displacement pressure and the ratio of drainage to imbibition displacement pressure have been proposed as qualitative indicators of preferential wettability of porous media. An increase in displacement pressure or in the ratio of drainage to imbibition displacement pressure signifies a tendency of the core to become more oil-wet. The above technique is applicable when oil-water capillary tests are made on native-state cores. However, most capillary pressure tests are either of the mercury injection or air-brine type, which provide little information concerning wettability.[81]

R. Resistivity Index Method

Formation resistivity obtained from electric logs can be used as a qualitative technique for wettability identification. Resistivity index is defined as the ratio of true formation resistivity to resistivity of the formation when 100% saturated with formation water. A high value of resistivity index indicates a low water saturation or a discontinuous water phase, which characterize an oil-wet system. A knowledge of the water saturation in the rock may yield sufficient information to make a judgement about rock wettability.

There is considerable uncertainty concerning the nature of the wettability characteristics of reservoir rocks *in situ*. Tests of wettability made on cores taken from reservoirs are not necessarily valid indicators of subsurface conditions, since the coring process itself may alter wettability. Cores cut in oil-base mud, for example, are often rendered entirely or partially preferentially oil-wet. Therefore special precautions must be observed during both coring and transporting to minimize the danger of altering the true wettability of the rock. In the absence of convincing evidence to the contrary (for example, abnormally high resistivity index) the assumption of preferential water wettability has been frequently used.[86]

VII. FACTORS INFLUENCING WETTABILITY EVALUATION

It has been suggested that four factors may influence the results of experimental determination of rock wettability.[87] One of these factors is core recovery and preservation. In the process of core recovery from a reservoir, heavy hydrocarbon components of crude oil become less soluble as the oil loses its associated solution gas (as a result of pressure reduction). The heavy hydrocarbon components can precipitate on the rock grains, leading to less water-wet or even oil-wet core behavior.[88-90] Drilling fluid containing surface-active materials may drastically change a core wettability, but it has been shown that bentonite and carboxymethyl cellulose have no observable effect on rock wettability when they are used in the coring fluid.[74] Weathering and contamination of cores during preservation and storage are also found to influence core wettabilities.[91] Strongly water-wet cores may become less water-wet as a result of air exposure, while cores with intermediate wettability show no significant change.[65] Oil-wet cores also may become water-wet upon exposure to air.[72] It has been suggested that alteration due to air exposure can be minimized and native-state wettability can be restored by incubation of the core in reservoir oil for two weeks at reservoir temperature.[28]

Crude oil is probably the best coring fluid for preserving wettability and maintaining native interstitial water saturation;[92] however, use of the wetting phase as a coring fluid may preserve the rock properties properly.[28] NaCl brine containing $CaCO_3$ powder with no other additives is considered a good fluid for cutting cores.[93] Care must be taken to avoid contamination of the coring fluid with air, sediments, etc. The use of crude oil as a coring fluid is likely to introduce a fire hazard into the coring operation, especially if a high API gravity oil is used.

Native state wettability of cores is obviously the most desirable condition, and the best technique for obtaining cores in this condition is by employing a pressure core barrel. The method allows cores to be cut and retrieved at reservoir pressure. At the surface, the cores are frozen, cut into sections, and sent to the laboratory.[94] Although early attempts at pressure coring met with limited success, recent developments indicate a success ratio of 80 to 90%.

Cores that have been cleaned, dried, and restored to some saturation and wettability condition are known as "restored state" cores.[48] This technique has been employed for many years and it is an established procedure; unfortunately, quite frequently, the cores are not restored to their native state and the use of these cores invalidates results obtained using sophisticated measurement techniques. Put very simply, restored state cores are not.

Factors that influence the core wettability evaluation include the laboratory core cleaning and preparation procedure. Mungan[92] states that the cleaning procedure neither changes the pore size distribution nor the quantity of kaolinite and illite in the core. He concludes that the change in fluid flow behavior is basically due to wettability alteration. Salathiel[39] reasons that the extraction of a core with strong solvents dissolves the strongly oil-wet surface coating of heavy organic molecules and thereby alters fluid displacement behavior of many fresh or preserved cores, as shown in Figure 23.[28]

Jenning's[95] results show a small but measurable change in the water-oil relative permeability ratio curve after toluene extraction of a variety of core samples from oil-bearing sandstones and limestones. The changes are not thought to be caused by significant changes in wettability. The results of Richardson et al.[91] show a higher rate of imbibition and a lower irreducible water saturation when East Texas Woodbine cores are extracted by hexane and methanol. Morgan and Gordon's[28] results show that the effect of cleaning procedure on core wettability may be minimized if reservoir fluids are used as testing fluids. Richardson et al.[91] believe a change in fluid flow behavior occurs as a result of repeated flooding of East Texas Woodbine cores. This change appears as a decrease in irreducible water saturation and as an increase in residual oil saturation. Further work is necessary for better understanding of this problem.

A third category of factors that influence core wettability evaluation is the testing condition. Stainless steel wettability can be altered by pressure increase in a methane-water system.[96] In spite of decrease in interfacial forces, the oil-water-solid system became more water-wet with temperature increases in a clean unconsolidated Houston sand and a natural unconsolidated California oil sand.[97] One explanation for the effect of temperature on displacement behavior is that polar components of the crude oil may not be adsorbed as readily on the grain surfaces of a rock at elevated temperature, so the flow behavior becomes more water-wet.[74,98]

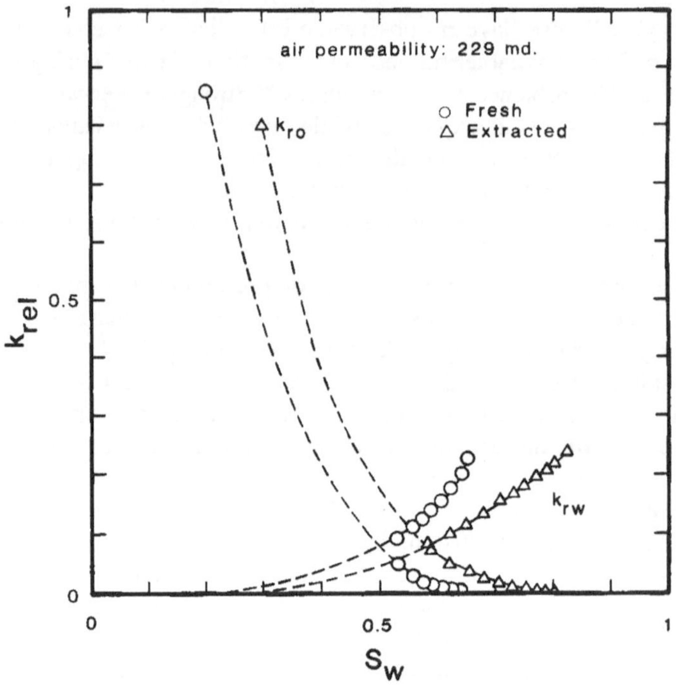

FIGURE 23. Comparison of relative permeability data from the same sample in fresh and extracted state.[28]

A fourth category of factors that influence the core wettability evaluation is the type of fluid used in the test. Carbonates are very sensitive to nitrogeneous surfactant compounds containing sulfur and oxygen.[41] Sandstones containing large percentages of silica possess acid type surfaces.[38,91] Crude oil containing normal paraffins are inert and inactive with regard to the surfaces of porous media, while napthene and aromatics are more active with porous surfaces. Heterocyclics and asphaltenes containing oxygen, nitrogen, sulfur, and metallic atoms are active with regard to the acid or basic sites. Reisberg and Doscher[40] have indicated that different crude oils probably have different proportions of these compounds which are believed to be responsible for the wettability characteristics of surfaces.

The critical gas saturation decreases and that for oil increases with increasing concentrations of polar substances.[99] Furthermore, increasing the concentration of polar compounds in oil causes the cumulative water production to increase and cumulative oil production to decrease in laboratory tests.

Oxidation of crude oil frequently appears to modify the wettability of porous media. The degree of modification depends on the amount of oxidizable polar compounds in contact with air and wettability may even be reversed.[98] Morgan and Gordon[28] and Cuiec[98] have investigated the effect of fluids and laboratory handling on relative permeability. Mungan[92] saturated an extracted core with reservoir fluid and let it sit at reservoir temperature for 6 days. He discovered that the measured relative permeability values were identical to those of freshly preserved cores; but when he used purified fluids in place of reservoir fluids a more water-wet condition in the core was developed, as indicated in Figure 24.

The initial fluid saturation in a core,[90] salinity alteration,[90] water alkalinity and hardness,[99] as well as the aging process[91] can influence the preferential wettability of a core. Wagner and Leach[56] have shown that the wettability of an oil- or intermediately wet sample of sandstone or carbonate can be changed to a more water-wet condition by the addition of chemicals such as hydrochloric acid, sodium hydroxide, and sodium chloride. They inves-

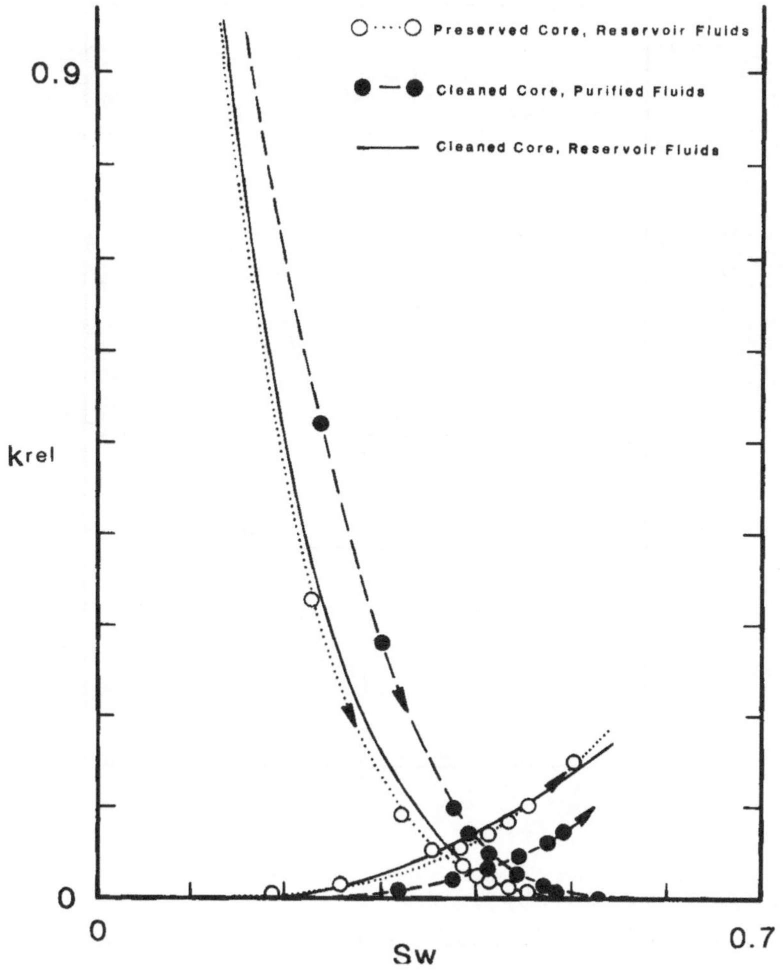

FIGURE 24. Effect of fluid and laboratory handling on relative permeability.[92]

tigated the influence of water pH on wettability of a quartz sample and used a n-octylamine treated synthetic oil to produce an oil-wet quartz surface. Their results indicated that lower pH solutions tend to produce water-wet surfaces under controlled salinity conditions. This effect is shown in Figure 25.

Bradley[100] has shown that a basic 5% NaCl solution spontaneously decreases the contact angle of oil-wet cores and as a result increases the amount of imbibition. These effects were reported to be most pronounced on cores of intermediate wettability. Morrow et al.,[63] Wagner and Leach,[56] and McCaffery and Mungan[101] have shown that wettability of typical reservoir rocks can be easily changed to any desired degree by adding polar compounds such as amines or carboxylic acids. Bradley[100] found that carboxylic acids such as stearic acid CH_3 $(CH_2)_{16}$ COOH at concentrations greater than 10^{-6} mol/ℓ altered the wettability of a water-dodecane-calcite system toward more oil-wetness and stearic acid with a concentration of approximately 5×10^{-3} mol/ℓ caused strongly oil-wet surfaces. He found that stearic acid caused no wettability alteration when quartz samples were used. Bradley found that amines such as octadecylamine CH_3 $(CH_2)_{17}$ NH_2 alter the wettability of both quartz and calcite toward oil wetness, especially at concentrations greater than 5×10^{-4} mol/ℓ. It should be noted that polar compounds which alter wettability of a given rock type may not alter the wettability of another rock type.

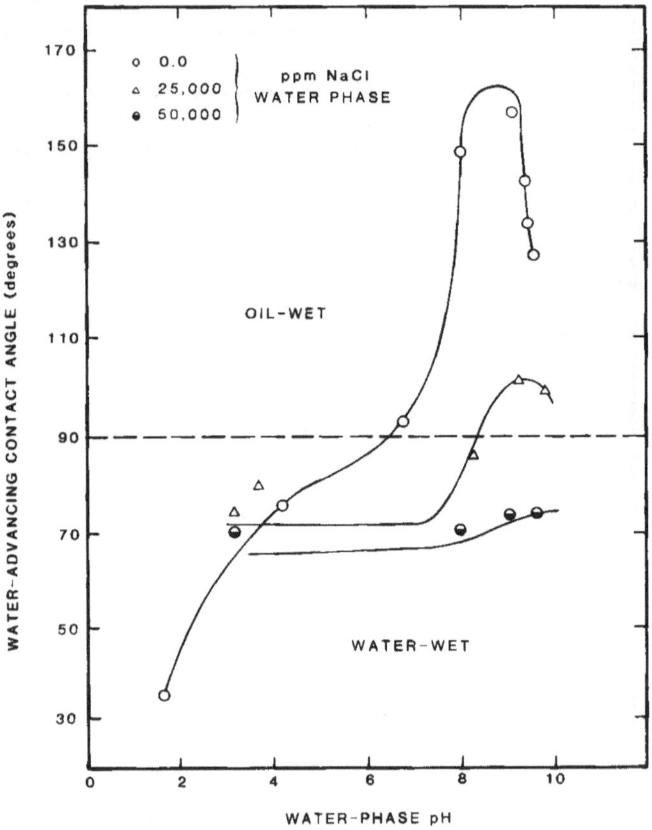

FIGURE 25. Contact angle as a function of pH.[56]

VIII. WETTABILITY INFLUENCE ON MULTIPHASE FLOW

The microscopic distribution of fluids in a porous medium is greatly influenced by the degree of rock preferential wettability. The fluid distribution in virgin reservoirs under strongly water-wet and strongly oil-wet conditions has been described by Pirson.[102] In a strongly water-wet reservoir, most of the water resides in dead-end pores, in small capillaries, and on the grain surface. In strongly oil-wet reservoirs, water is in the center of the large pores as discontinuous droplets, while oil coats the surfaces of the grains and occupies the smaller capillaries.

Under strongly water-wet conditions the effective permeability to the nonwetting phase at irreducible water saturation is approximately equal to the absolute permeability of the rock. On the other hand, in strongly oil-wet systems, the effective permeability to oil at irreducible water saturation is greatly reduced by the water droplets in the larger pores. Raza et al.[66] stated that in some oil-wet reservoirs, water occupies some of the finer pores and is trapped as droplets in the larger ones. Raza et al. analyzed the displacement of oil by advancing water and the trapping of the residual oil as shown in Figure 26.

In strongly water-wet reservoirs, water traps oil in the larger pores as it advances along the walls of the pore, while in strongly oil-wet reservoirs, water moves in large pores and oil is trapped close to the walls of the pores.[66]

The petroleum industry has long recognized that the wettability of reservoir rock has an important effect on the multiphase flow of oil, water, and gas through the reservoir. API Project 27 at the University of Michigan was initiated in 1927 to study this problem. The

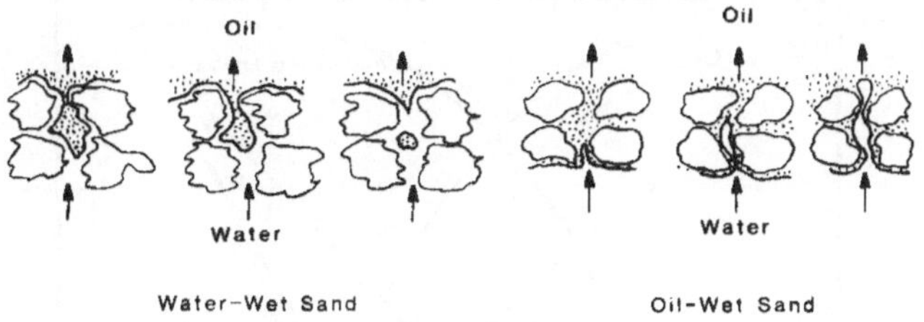

FIGURE 26. The trapping process of oil by advancing water.[66]

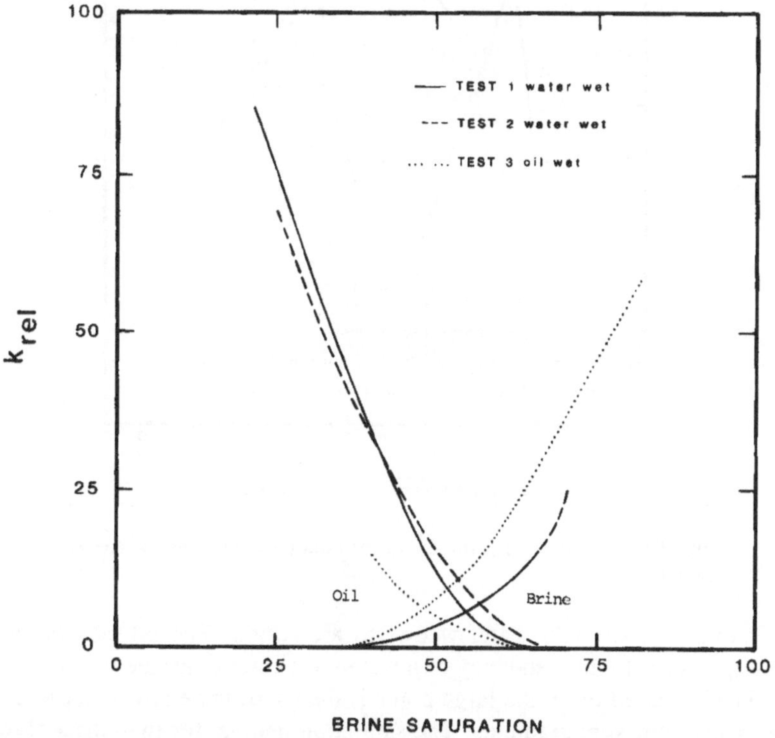

FIGURE 27. Effect of wettability on flow behavior.[12]

dissymmetry of relative permeability curves is attributed largely to the preferential wettability of reservoir rock.[79,95,103] As illustrated by Figure 27, Geffen et al.[12] and Donaldson and Thomas[104] have shown the effect of fluid distributions brought about by rock preferential wettability on the relative permeability-saturation relationship. As the degree of rock preferential wettability for water decreases, the oil relative permeability at a given saturation decreases while the water relative permeability increases.

Schneider and Owens[84] recognized the fact that rock type appears to have less influence on flow relationships than does rock wetting preference. However, this may not be the case for heterogeneous rocks or mixed wettability systems. Owens and Archer[11] also confirmed the importance of preferential wettability on multiphase flow in porous media.

Some investigators[90] have found that relative permeability becomes progressively less favorable to oil production as a rock becomes less water-wet. The residual oil saturation increases as a rock becomes less water-wet. Others have shown that weakly water-wet cores have more favorable relative permeability curves and lower residual oil saturations than strongly water- or oil-wet rocks. Conceptually, this latter behavior seems reasonable since

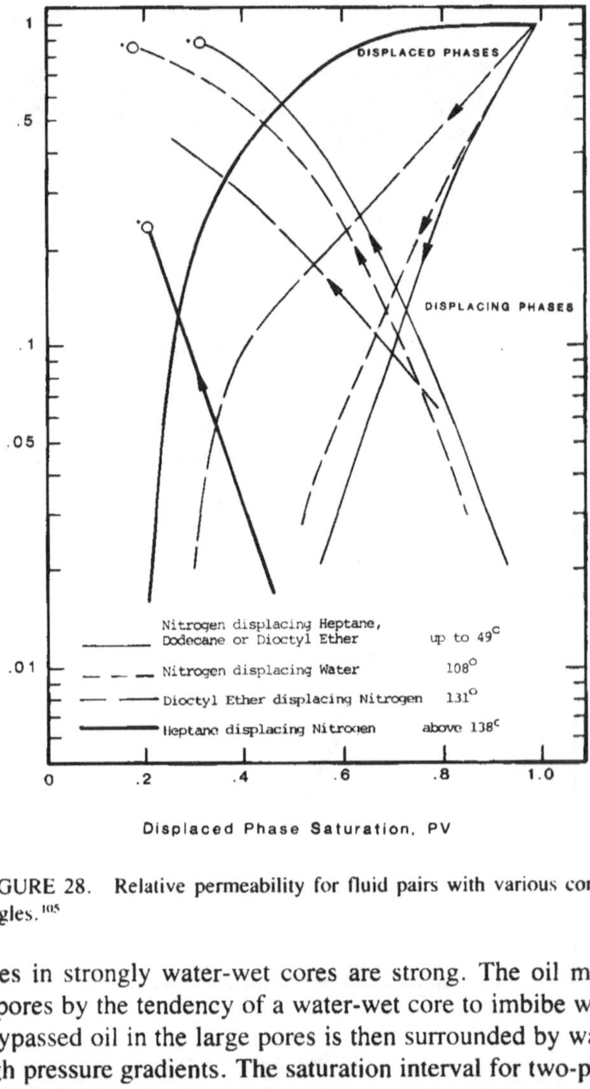

FIGURE 28. Relative permeability for fluid pairs with various contact angles.[105]

the capillary forces in strongly water-wet cores are strong. The oil may be bypassed and trapped in larger pores by the tendency of a water-wet core to imbibe water into the smaller capillaries. The bypassed oil in the large pores is then surrounded by water and is immobile except at very high pressure gradients. The saturation interval for two-phase flow under this condition is probably short.

As the capillary forces are reduced by reduction in preferential water-wettability of a rock, the tendency toward rapid imbibitional trapping of oil in large pores by movement of water through small pores should also diminish. The zone of two-phase flow should become broader and oil displacement to a lower residual saturation should be possible. If other factors remain constant, higher flow rates and lower interfacial tensions are conducive to higher oil recovery; these are changes that diminish the ratio of capillary forces to viscous forces.

Stegemeier and Jensen[37] and McCaffery and Bennion[105] reported that wettability alterations over a relatively wide range produce a negligible effect on the relative permeability curve, as shown by Figure 28. However, other workers did not confirm this finding. Treiber et al.[62] found that relatively small variations in wettability produce considerable effects on the relative permeability curve. Figure 29 shows the effect of contact angles on relative permeability curves for a Torpedo sandstone.

IX. EFFECTS OF SATURATION HISTORY

The relative permeability-saturation relation is not a unique function of saturation for a given core, but is subject to hysteresis for porous systems with strong wetting properties.

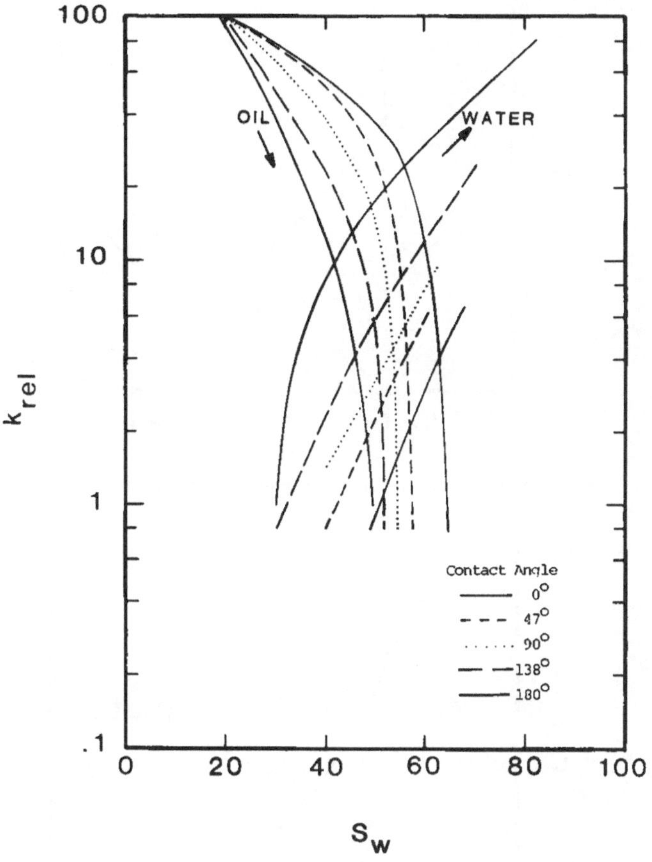

FIGURE 29. Imbibition relative permeability with various contact angles.[62]

That is, the relative permeability of a porous medium to a fluid at a given saturation depends on whether that saturation is obtained by approaching it from a higher value or a lower one. In a displacement process where the wetting-phase saturation is approached from a lower value, the resulting relative permeability curve is referred to as an imbibition curve (an increase in the wetting phase). Examples of imbibition processes are the injection of water during waterflooding and coring a water-wet rock with a water-base mud. On the other hand, in a displacement process where the wetting phase saturation is approached from a higher value, the resulting relative permeability curve is referred to as a drainage curve. Examples of drainage processes are the displacement of oil by expansion during primary depletion of a reservoir and the accumulation of hydrocarbons in oil and gas reservoirs; another example would be waterflooding an oil-wet reservoir.

Geffen et al.,[12] Osoba et al.,[13] Levine,[106] Josendal et al.,[107] Terwilliger et al.,[108] and Coley et al.[81] described the hysteresis phenomenon and verified that both water-oil and gas-oil relative permeability ratio curves as well as individual wetting and nonwetting phase relative permeability of both sandstone and carbonate formations may exhibit hysteresis.[11,22,107,109] In a two-phase system, hysteresis is more prominent in relative permeability to the nonwetting phase than in relative permeability to the wetting phase.[10,110] The hysteresis in wetting-phase relative permeability is believed to be very small and thus, sometimes difficult to distinguish from normal experimental error, as indicated in Figure 30.

The drainage curve shown in Figure 30 is a primary drainage curve which is applicable only when drainage occurs before imbibition. When a drainage process occurs after imbibition, a secondary drainage curve exists, as shown in Figure 31.

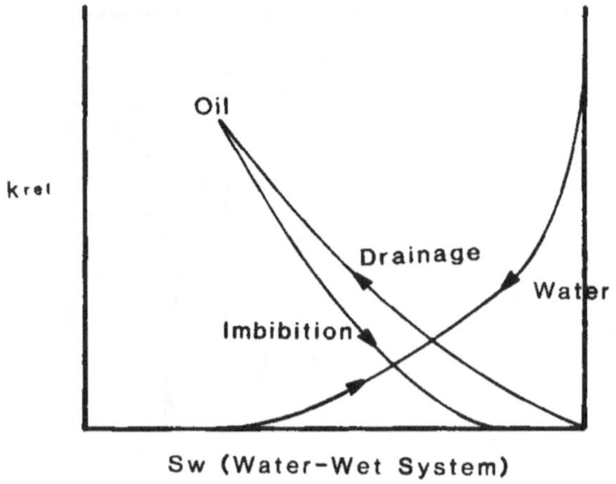

FIGURE 30. Primary drainage relative permeability curve.

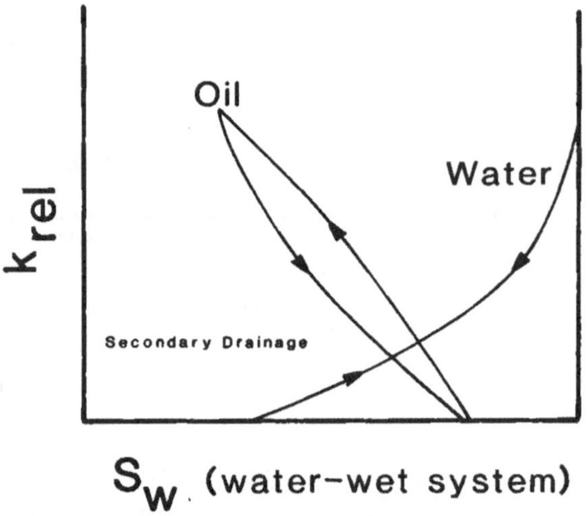

FIGURE 31. Secondary drainage curve: end-point flow reversal.

These curves describe relative permeability when the flow reversal occurs at one of the saturation end points. The effect of flow reversal at an intermediate saturation value is illustrated by Figure 32.

As shown in Figures 30 and 31, the water (wetting phase) relative permeability curve is essentially the same in strongly water-wet rock for both drainage and imbibition processes.[11] However, at a given saturation, the nonwetting phase relative permeability of a consolidated rock is usually less for an imbibition cycle than for a drainage cycle.[12,13,22,106] For an unconsolidated rock, the nonwetting phase relative permeability in an imbibtion cycle is usually greater than the corresponding nonwetting phase relative permeability in a drainage cycle. Naar et al.[22] reported that relative permeability relationships for poorly consolidated formations tend to resemble those for unconsolidated formations.

Figure 33 shows the imbibition and drainage relative permeabilities of a consolidated rock. It can be seen that the residual nonwetting phase saturation is much greater for imbibition than for drainage. That is, the nonwetting phase loses its mobility at a higher saturation in imbibition than it does in drainage. Figure 34 shows that the imbibition cycle k_{ro} may lie above k_{ro} on the drainage cycle for some systems. This relationship probably is not typical of petroleum reservoirs.

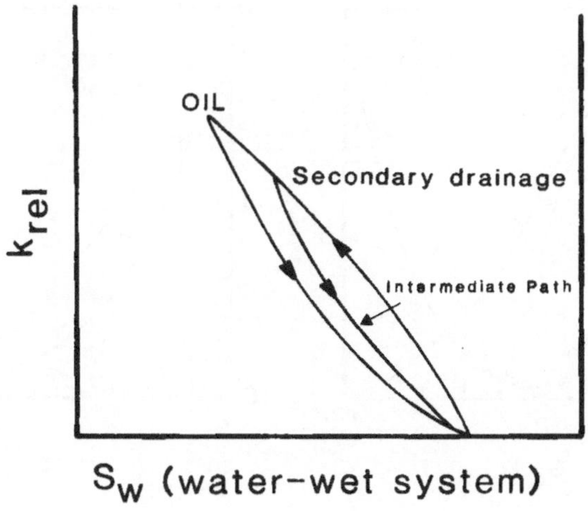

FIGURE 32. Secondary drainage curve: intermediate flow reversal.

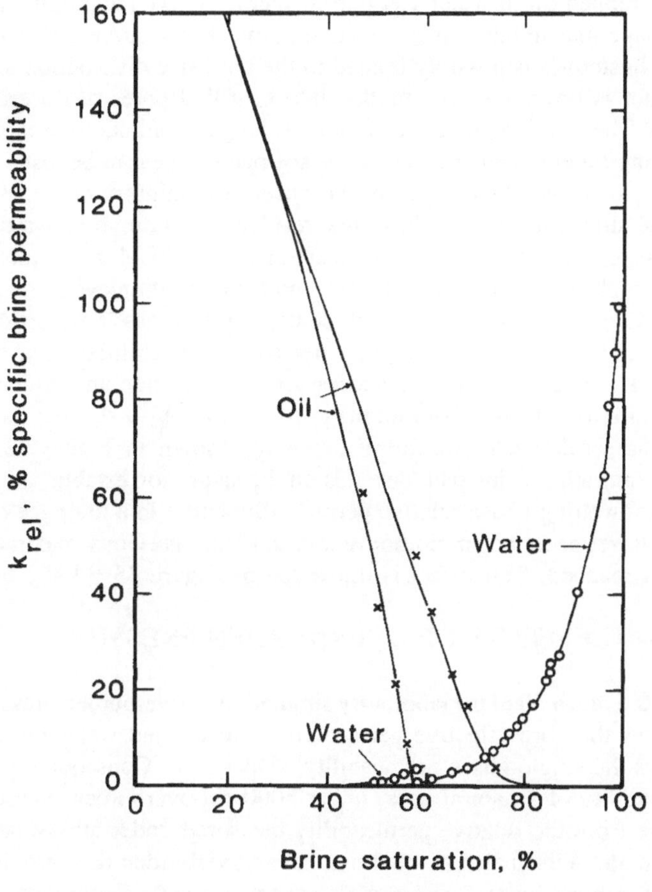

FIGURE 33. Oil-water flow characteristics of a consolidated rock.[12]

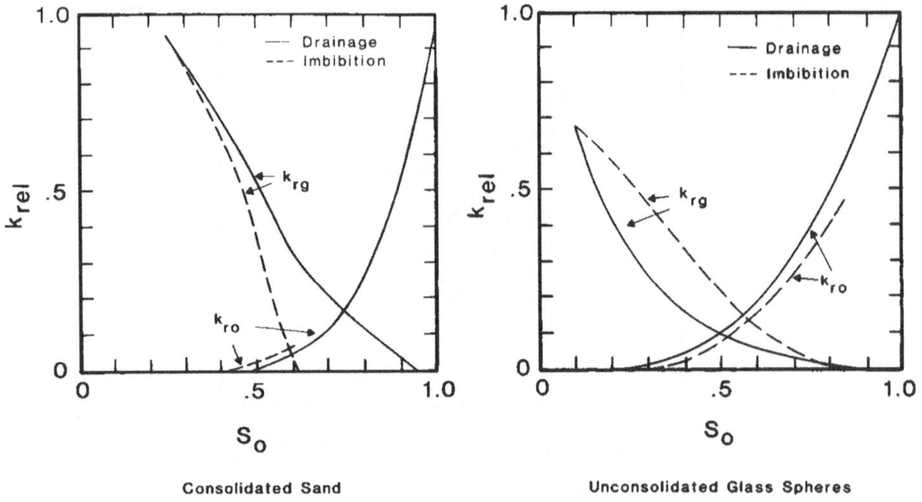

FIGURE 34. Relative permeability curves for consolidated sands and unconsolidated glass spheres.[22]

The amount of trapped oil in water-wet porous media is given approximately by the area between the drainage and imbibition oil relative permeability curves.[112] It is believed that the occurrence of hysteresis is possibly related to the pore size distribution and cementation of a rock. As water is progressively imbibed into oil-filled pores of different sizes, oil is ejected from them. The ejection process continues as long as continuous escape paths through pores still containing oil are available. These escape paths appear to be lost at oil saturations which greatly exceed those which occur at the onset of continuity of a nonwetting phase, (e.g., gas) on the drainage cycle. Thus, the residual oil saturation which results from waterflooding a water-wet rock is much greater than the critical gas saturation that characterizes the same rock. Apparently oil is trapped on the imbibition cycle. A similar behavior is observed if a preferentially water-wet rock containing free gas is waterflooded.

The imbibition and drainage wetting-phase relative permeabilities of a consolidated or unconsolidated rock are retraced under a succession of imbibition and drainage cycles; in a reversal of the saturation change from drainage to imbibition, a distinct path is traced by the nonwetting phase relative permeability curve (as shown in Figure 32) to a residual nonwetting phase saturation. This path depends on the saturation established in the drainage cycle. Also, the nonwetting phase relative permeability curve in a drainage cycle following an imbibition cycle retraces the imbibition curve until the previous maximum nonwetting phase saturation is reached. This effect is illustrated by Figure 35.[22,113]

X. EFFECTS OF OVERBURDEN PRESSURE

Wilson[114] reported that a 5000 psi laboratory simulation of overburden pressure at reservoir temperature reduces the core effective permeabilities to oil and water by about the same extent as it reduces the single-phase permeability of that core. Consequently, the water and oil relative permeability of a natural core, under 5000 psi overburden pressure, show only a moderate change from the relative permeability measured under atmospheric conditions, as shown in Figure 36. Wilson also pointed out that an overburden pressure that can produce over 5% reduction in porosity of a core can also produce a sufficiently large change in pore size distribution to affect the relative permeability of the core.

In contrast to the work of Wilson, Fatt and Barrett[115] concluded that variation of rock overburden pressures in the range of 3000 psi does not produce any change on gas relative permeability in a sandstone gas-oil system. Figure 37 shows the gas relative permeability

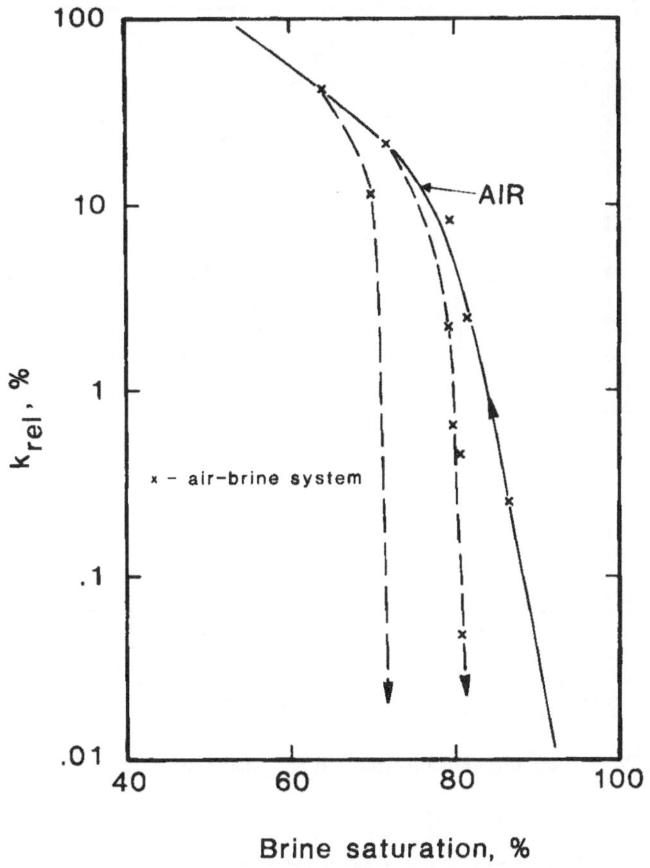

FIGURE 35. Air flow behavior in two-phase systems, Nellie Bly sandstone.[12]

with and without the laboratory simulation of overburden pressure. Similar results were reported by Thomas and Ward[116] for a gas-oil system in a low permeability rock. Geffen et al.[12] have shown that the residual gas saturation in a liquid-gas system, under atmospheric conditions, is similar to the resisdual gas saturation measured under a 5000 psi laboratory simulation of overburden pressure. Merliss et al.[117] concluded that the effect of overburden pressure on relative permeability was primarily due to changes in interfacial tension.

XI. EFFECTS OF POROSITY AND PERMEABILITY

Wyckoff and Botset[3] as well as Leverett and Lewis[8] investigated the influences of porosity and absolute permeability on relative permeability and found them to be insignificant. Dunlap[118] used unconsolidated sand packs having permeabilities of 3.0, 4.5, and 8.0 D and found no indication that the relative permeability-saturation relationship is a function of specific permeability of the sand. Stewart et al.[119] found that variations in permeabilities ranging from 8.5 to 300 mD and porosities from 15 to 22% in limestone cores with intergranular porosity, caused relative permeability curves to shift up to a maximum of 2% of gas saturation. These investigators employed a solution gas drive, gas-oil relative permeability measurement technique in their study. They also reported the relative permeability curves to shift up to a maximum of 4% of gas saturation when fractured limestone cores of various porosities and permeabilities were employed.

Botset[21] found that absolute permeabilities ranging from 17 to 260 D had negligible effects

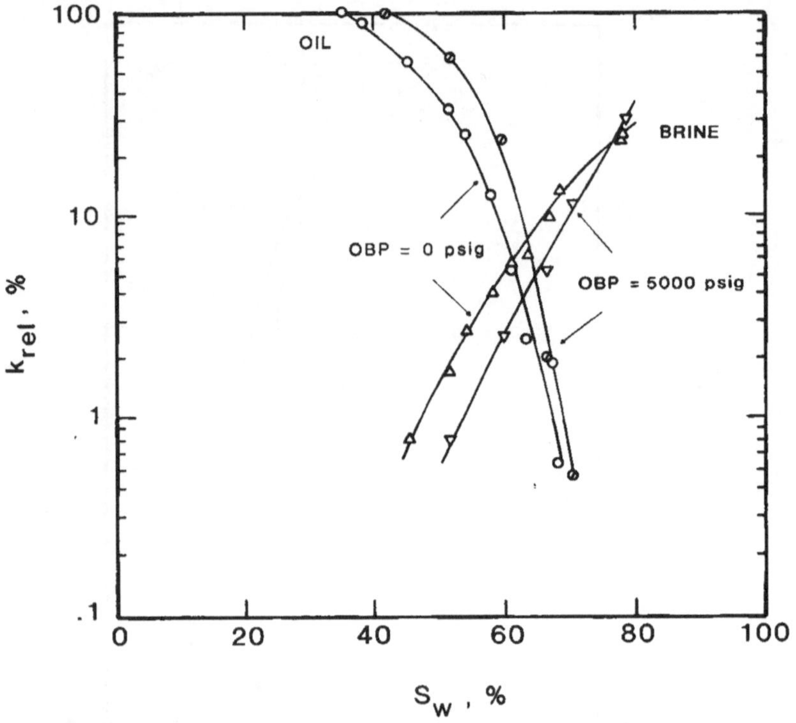

FIGURE 36. Effect of overburden pressure on relative permeability of an oil-brine system.[114]

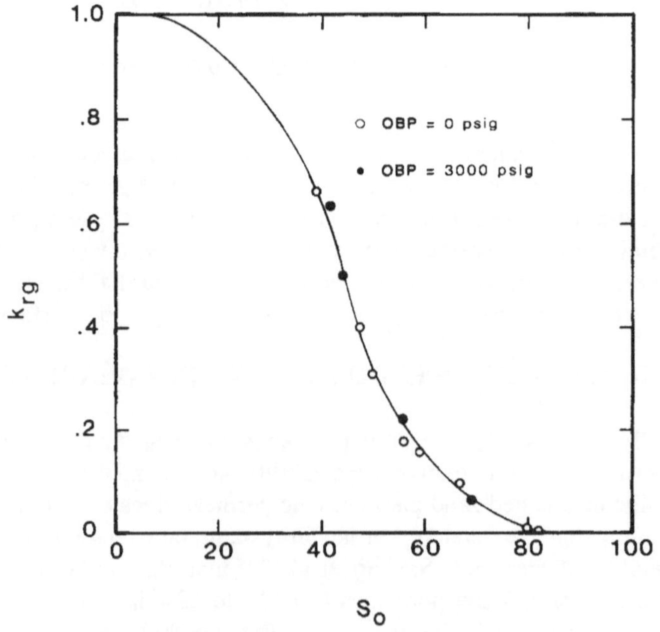

FIGURE 37. Effect of overburden pressure on gas relative permeability.[115]

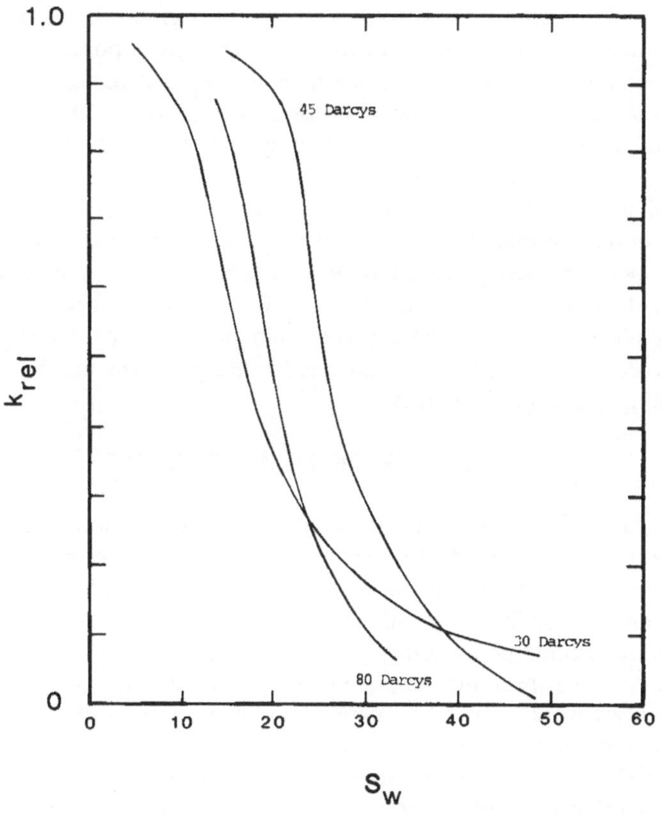

FIGURE 38. Effect of absolute permeability on relative permeability.[30]

on the gas-liquid relative permeability-saturation relationship of a consolidated Nichols Buff sandstone. Botset's results were in agreement with the findings of Leverett,[4] who used sands with permeabilities ranging from 1.04 to 6.80 D.

Morgan and Gordon[28] conducted tests on four sandstone samples from a reservoir rock with permeabilities ranging from 109 to 273 mD. No clear effect of permeability on oil-water relative permeability curves was observed. Crowell et al.[30] studied four different sands with absolute permeabilities ranging from 3.0 to 8.0 mD and found no correlation between absolute permeability and gas relative permeability in a water-gas system as shown in Figure 38.

Keelan[120] observed satisfactory correlations of sandstone air permeability corrected for slippage and the irreducible water saturations as well as end-point relative permeability values of gas-water systems. Leas et al.[121] noted a correlation between absolute permeability and gas relative permeability in particular cases, but believed this relationship not to be true in general.

Felsenthal[122] tested 300 sandstone cores and noted that the gas-oil relative permeability curves became less steep as specific permeability increased. This trend was also reported by McCord.[123] In Felsenthal's paper an effect of porosity on gas-oil relative permeability ratio was also noted. This effect was not generally discernible in the study of relative permeability data for a given reservoir but became apparent when data for sandstone reservoirs of similar lithology but differing average porosity were compared. For example, a definite trend was observed in a comparison of argillaceous and/or calcareous sandstones from 11 reservoirs ranging in average porosity from 14 to 28%, indicating that for a given permeability, the gas-oil relative permeability ratio curves became less favorable, (i.e., k_g/k_o increased) as porosity increased. A similar trend was observed for a group of clean sandstones from five reservoirs ranging in porosity from 15 to 21%. For a given porosity and permeability, comparatively clean sandstones gave more favorable gas-oil relative permeability ratio curves than argillaceous and/or calcareous sandstones or chert reservoirs. The least favorable gas-oil relative permeability ratio curves were for conglomerates, shaly sandstones, and sandstones containing carbonate inclusions. Felsenthal then classified sandstones in three categories and found a correlation of gas-oil relative permeability ratio for each class. The parameters used in the correlation were porosity, permeability, and sandstone type, which are all related to pore geometry. On the other hand, pore geometry may be characterized by the pore size distribution and Felsenthal found a correlation between gas-oil relative permeability ratio and pore size distribution. He found that the more favorable gas-oil relative permeability ratio curves were generally associated with a pore size distribution curve having a sharp peak among the large pore sizes.

XII. EFFECTS OF TEMPERATURE

Several early studies[124-128] indicated that irreducible water saturation increased with increasing temperature and that residual oil saturation decreased with increasing temperature; all of these studies employed a dynamic displacement process. Difficulties in evaluating these results include possible wettability changes due to the core-cleaning procedure,[126] possible changes in absolute permeability, and clay migration.[124,127,128]

Steady-state relative permeability measurements by Lo and Mungan[129] indicated that the relative permeabilities were temperature-dependent when using white oils, but were unaffected by temperature changes when using tetradecane; this finding agrees with the results of Edmondson.[124] Other variations in results have been attributed to viscosity ratio. Sufi et al.[130,131] pointed out that some of the previous results may have significant error due to the difficulty in measuring relative permeabilities at elevated temperatures and suggested that temperature effects possibly result from a combination of measurement difficulties and laboratory-scaling phenomena, (i.e., end effects in short cores).

Miller and Ramey[132] performed dynamic displacement experiments at elevated temperatures on unconsolidated sand packs and a Berea core. Their results indicated that changes in temperature do not cause relative permeability changes, but that changes in the flow capacity at elevated temperatures are due to clay interactions, change in pore structure, etc. The only change that they observed was an increase in oil relative permeability at irreducible water saturation and this parameter is relatively unimportant for predicting two-phase flow behavior. In measuring steam-water relative permeabilities, Counsil[133] and Chen et al.[167] also noted the absence of temperature effects.

XIII. EFFECTS OF INTERFACIAL TENSION AND DENSITY

The interfacial forces at fluid-fluid and fluid-solid interfaces are responsible for retention of residual saturation in porous media. Wyckoff and Botset[3] and Leverett[4] described a small but definite effect of interfacial tension within the range of 27 to 72 dyne/cm on relative permeability. (See Figure 39.) Lefebvre du Prey[103] also identified the interfacial tension of fluids in a consolidated sample as a factor influencing the relative permeability and residual saturation values. Crowell et al.[30] found that a reduction in interfacial tension of a water-air system produced an increase in gas recovery and a decrease in residual gas saturation.

Muskat[134] discounted the possibility that the interfacial tension within the range of 27 to 72 dyne/cm can influence relative permeability. Owens and Archer[11] concluded that interfacial tension has no influence on either the water-oil relative permeability of a water-wet core or the gas-oil relative permeability of an oil-wet core. They found that water relative

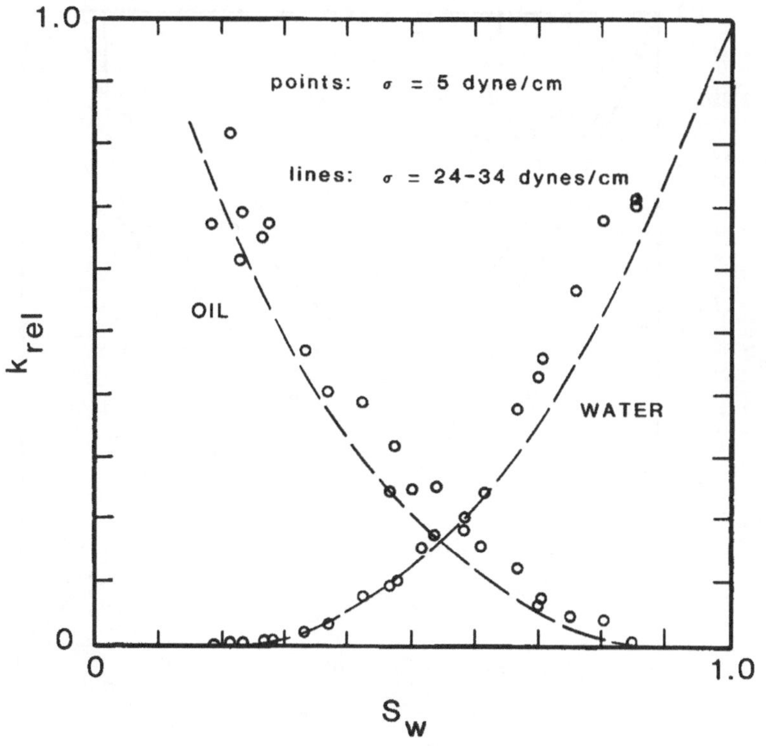

FIGURE 39. Effect of interfacial tension on relative permeability.[4]

permeability of the water-wet core and oil relative permeability of the oil-wet core were coincident.

Moore and Slobod[67] reported a reduction in waterflood residual oil saturation of a water-wet core at lower values of interfacial tension. Pirson[102] stated that drainage relative permeability is independent of the interfacial tension, but imbibition relative permeability is sensitive to interfacial tension. Bardon and Longeron[135] found that a reduction in interfacial tension reduced oil relative permeability at constant gas saturation in an oil-gas drainage cycle of the Fontainebleau formation. (See Figure 40.) The effect of liquid density on relative permeability has been found to be insignificant.[3,12]

XIV. EFFECTS OF VISCOSITY

Leverett et al.[4,8] investigated the effect of viscosity variation of an oil-water mixture on relative permeability of artificially compacted sands with 41% porosity and 3.2 to 6.8 D of absolute permeability. He found no systematic variation in relative permeability when the oil viscosity was varied from 0.31 cp (hexane) to 76.5 cp (lubricating oil) and the water phase viscosity was varied from 0.85 to 32.2 cp. Viscosity ratios employed in the study ranged from 0.057 to 90. The experiments of Leverett et al. were performed under steady-state flow at low pressure gradients. Figures 41 and 42 show the effect of viscosity ratio variation on water and oil relative permeability curves.

Wyckoff and Botset[3] found that moderate variations in viscosities of the fluid phases in unconsolidated sand packs with permeabilities ranging from 3.2 to 6.0 D failed to produce any change in the relative permeability values. In their experiment a mixture of water and carbon dioxide was employed and water viscosity was adjusted between 0.9 and 3.4 cp by addition of a sugar solution to the water.

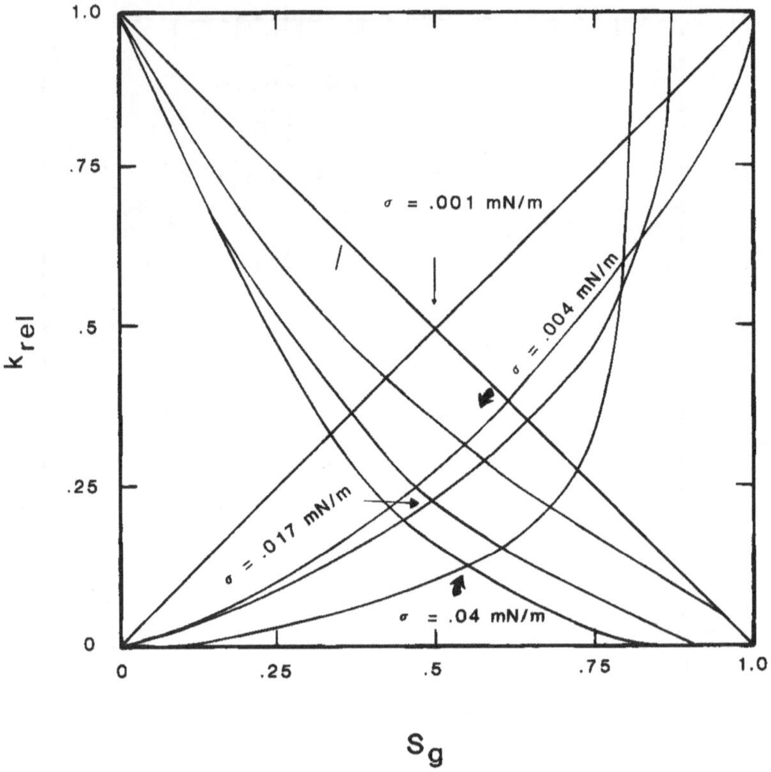

FIGURE 40. Effect of low interfacial tensions on gas-oil relative permeability.[135]

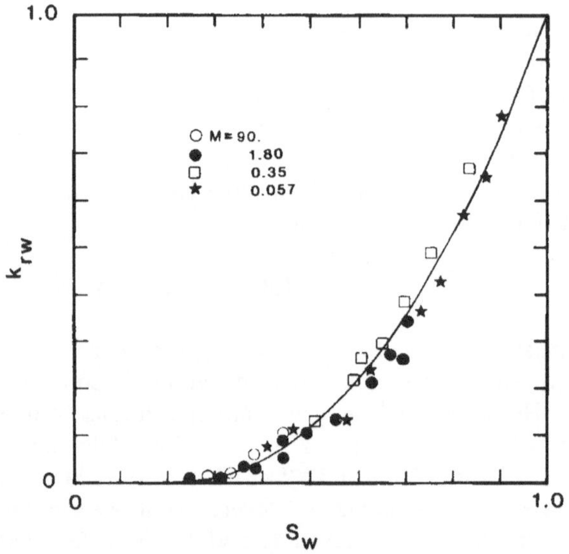

FIGURE 41. Effect of viscosity ratio (M) on water relative permeability.[4]

Richardson[136] found that the water-oil relative permeability ratio is independent of fluid viscosity where the oil viscosity varied from 1.8 to 151 cp (see Figure 43). Johnson et al.[137] confirmed these results for displaced/displacing viscosity ratios up to 37. Levine[138] found

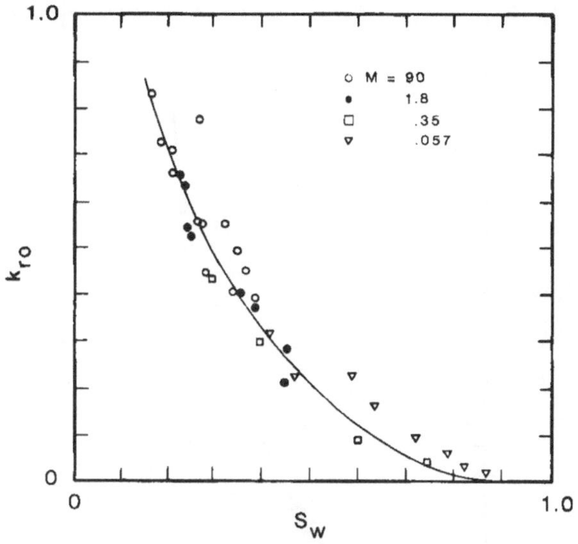

FIGURE 42. Effect of viscosity ratio (M) on oil relative permeability.[4]

that the relative permeability of a sandstone sample was independent of viscosity ratio in the range of 1.92 to 22.6. Craig[139] reported that the gas-oil relative permeability ratio of a Nellie Bly sandstone sample with 824 mD permeability and 28.1% porosity showed no significant variation with oil viscosities in the range of 1.4 to 125 cp. Results of this study are illustrated by Figure 44.

Sandberg et al.[140] found that oil and water relative permeabilities of a uniformly saturated core are independent of the oil viscosity in the range of 0.398 to 1.683 cp. Donaldson et al.[141] and Geffen et al.[142] also concluded that relative permeability is independent of viscosity as long as the core wettability is preserved. Wilson[114] found that a 5000 psi fluid pressure which caused kerosene viscosity to increase from 1.7 to 2.7 cp and water viscosity to increase by 1% did not produce any significant effect on water and oil relative permeability values. Muskat et al.[27] reported that the effect of viscosity on relative permeability of an unconsolidated sand was very small and within the limits of experimental accuracy.

Krutter and Day[143] used methane and air as the nonwetting phase in a two-phase system of oil and gas. The gas was injected into cores saturated with oils with viscosities ranging from 2 to 100 cP. They found that the air relative permeability values were slightly less than those for methane.

Saraf and Fatt[10] applied Darcy's law to each of the phases of a multi-phase system and concluded that relative permeability is independent of viscosity. The Saraf and Fatt equation is based on the assumption that different phases flow in different capillaries and do not come in contact with each other.

Yuster,[6] however, concluded that relative permeability values for the systems he studied were markedly influenced by variation in viscosity ratio, increasing with an increase of the ratio. This conclusion was later supported by the work of Morse et al.[144] Odeh[145] expanded Yuster's work and concluded that the nonwetting phase relative permeability increases with an increase in viscosity ratio. He found that the magnitude of the effect on relative permeability decreases with increase in single-phase permeability. Odeh found that the deviation in nonwetting phase relative permeability is increased as the nonwetting phase saturation is increased, with the deviation reaching a maximum at the nonwetting phase residual saturation. He also concluded that the wetting-phase relative permeability is not affected by variation in viscosity ratios. Figure 45 shows the effect of viscosity ratio variation in the range of 0.5

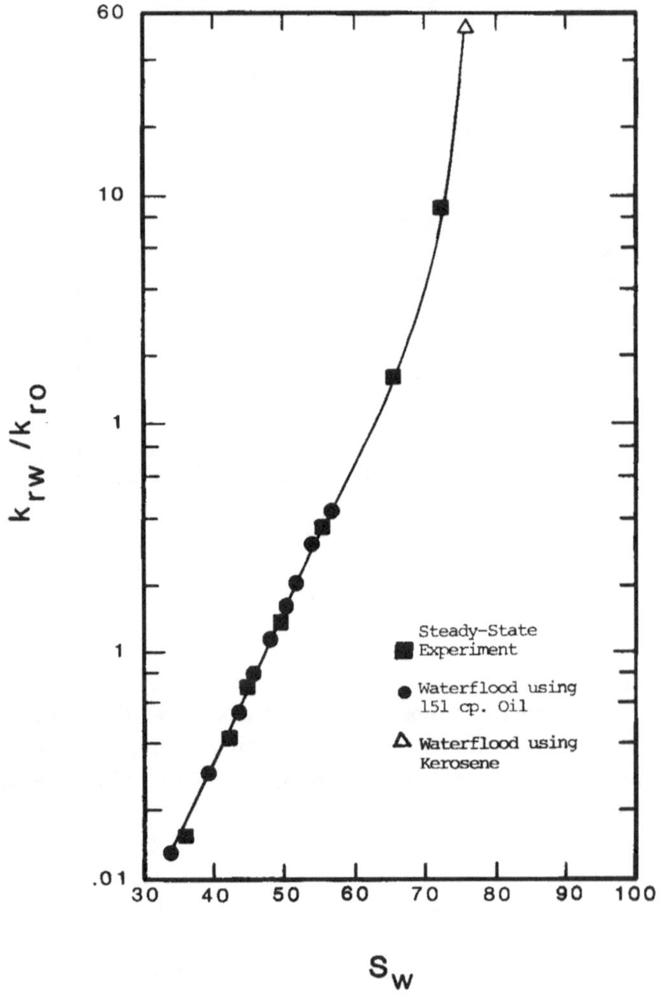

FIGURE 43. Comparison of steady-state results with flooding performance.[136]

to 74.5 on water and oil relative permeability curves. Odeh stated that the effect of viscosity ratio on relative permeability could be ignored for samples with single-phase permeabilities greater than 1D. Yuster's and Odeh's results have been criticized by other investigators.[146]

Downie and Crane[147] reported that oil viscosity could influence the oil effective permeability of some rocks. Later, they qualified their statement by saying that once an increased relative permeability is obtained by employment of high viscosity oil, it may not be lost by replacing this oil with one of a lower viscosity. They explained this phenomenon qualitatively in terms of the movement of colloidal particles at oil-water interfaces.

Hassler et al.[1] found that lower gas relative permeability values were associated with lower oil viscosity in a Bradford sand. However, they expressed doubt that the variation in relative permeability could be described by a single factor varying with oil viscosity.

Pirson[102] stated that the importance of the effect of viscosity ratio on the imbibition nonwetting phase relative permeability is of second-order magnitude. Ehrlich and Crane[148] concluded that the imbibition and drainage relative permeabilities, under a steady condition of flow, are independent of viscosity ratio. However, they found that the irreducible wetting-phase saturation following a steady-state drainage, when the interfacial effect predominated

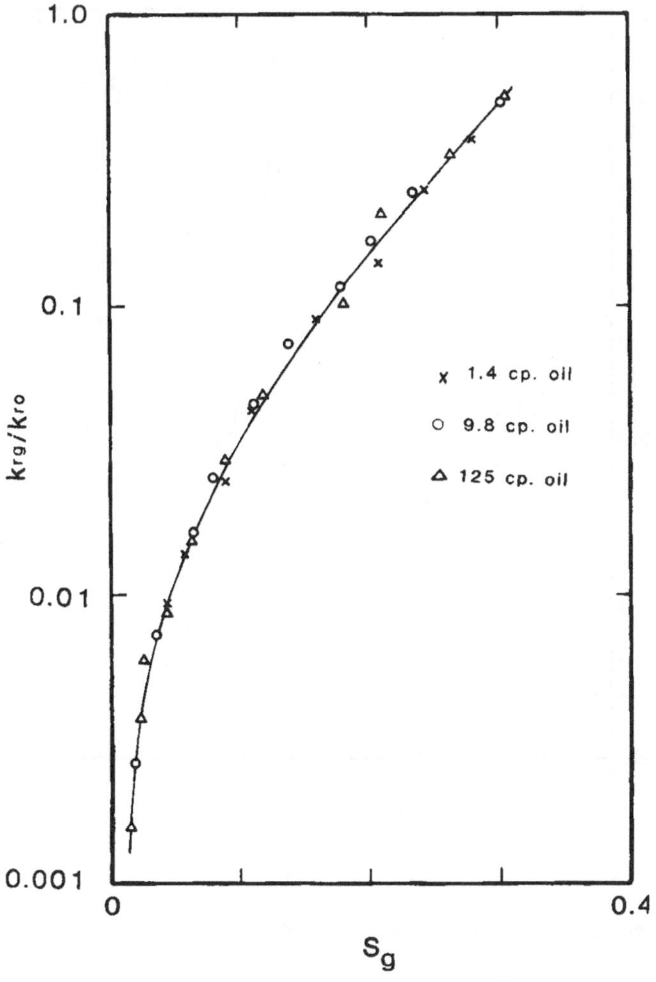

FIGURE 44. Relative permeability ratios for Nellie Bly sandstone.[139]

over viscous and gravitational effects, decreases with an increase in the ratio of nonwetting to wetting-phase viscosities.

McCaffery[59] reported that in strongly wetted systems, the imbibition and drainage relative permeabilities are independent of the viscous forces. He concluded that even though the relative permeability to a phase might be influenced by viscosity variation of that phase, the relative permeability ratio is independent of viscosity.

Perkins[149] concluded that flow in a porous body is governed by relative permeability and viscosity ratio when the ratio of capillary pressure to the applied pressure is negligible. Pickell et al.[150] concluded that only a large variation in viscous forces could have any appreciable effect on residual oil saturation. Several authors [4,67,151-153] recognized that the wetting and the nonwetting phase relative permeability might be significantly affected by the ratio of capillary to viscous forces, $\sigma\cos\theta/\mu v$, where σ represents interfacial tension expressed as dynes per centimeter; θ represents contact angle; μ represents viscosity expressed as cp; and v represents fluid velocity expressed as centimeters per second. Lefebvre du Prey[154] made a systematic study of the effect of this ratio on relative permeability by simultaneously varying the interfacial tension, viscosity, and velocity. He found that relative permeability decreases as the ratio $\sigma\cos\theta/\mu v$ increases. He also concluded that the relative permeability curve is influenced by the viscosity ratio when the wetting phase is displaced

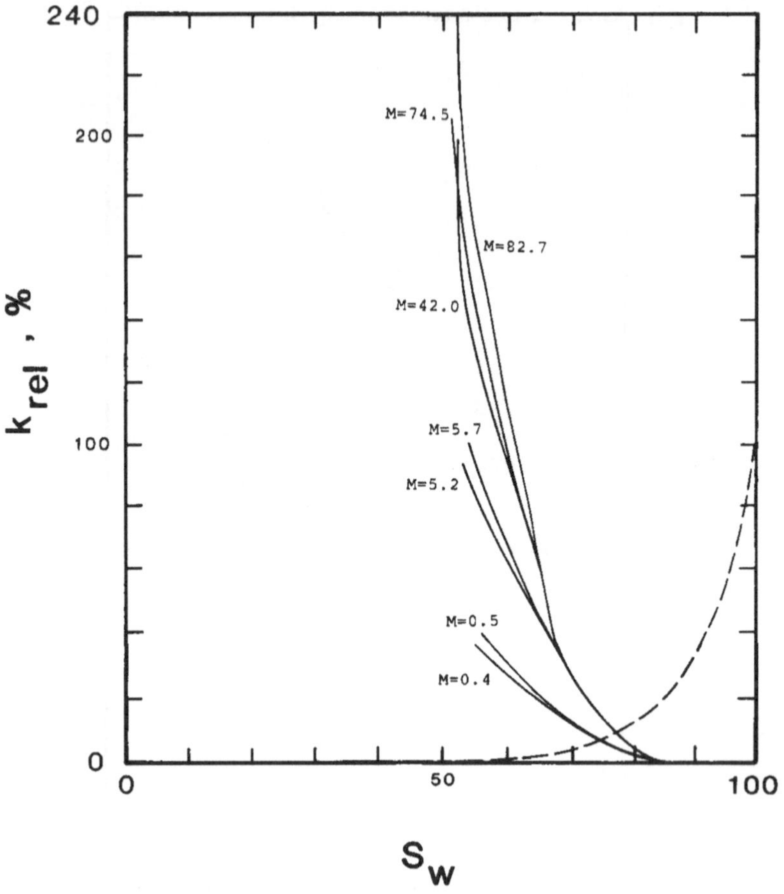

FIGURE 45. Effect of viscosity ratio (M) on relative permeability.[145]

by the nonwetting phase. Bardon and Longeron[135] found that in some gas-oil systems, the drainage relative permeability and residual oil saturation are strongly affected by the $\mu v/\sigma$ ratio.

An assumption that the relative permeability values are independent of viscosity implies that the system can be represented by a bundle of parallel, noninterconnecting capillary tubes, each of which is filled with either the wetting or the nonwetting phase alone. Thus, the nonwetting phase flows through the larger channels while the wetting phase flows through the smaller capillaries. However, this model probably does not completely represent the conditions in porous media. An alternative model is the simultaneous flow of two immiscible fluid phases in larger capillaries.

A flow picture more compatible with the present knowledge of fluid behavior is a combination of the two models described above, with one dominating over the other depending primarily on wettability. Odeh[7] believed that the fluid phases did not flow in separate capillaries of porous media as Leverett postulated and further stated that the wetting phase moves microscopically in a sort of sliding motion imparted to it by the shear force caused by motion of the nonwetting phase. From this model he concluded that a decrease in interstitial wetting-phase saturation can be developed as a result of an increase in viscosity, thereby affecting the relative permeability values.

In view of the diverse opinions which have been expressed by various investigators concerning the influence of viscosity on relative permeability, it seems best to conduct

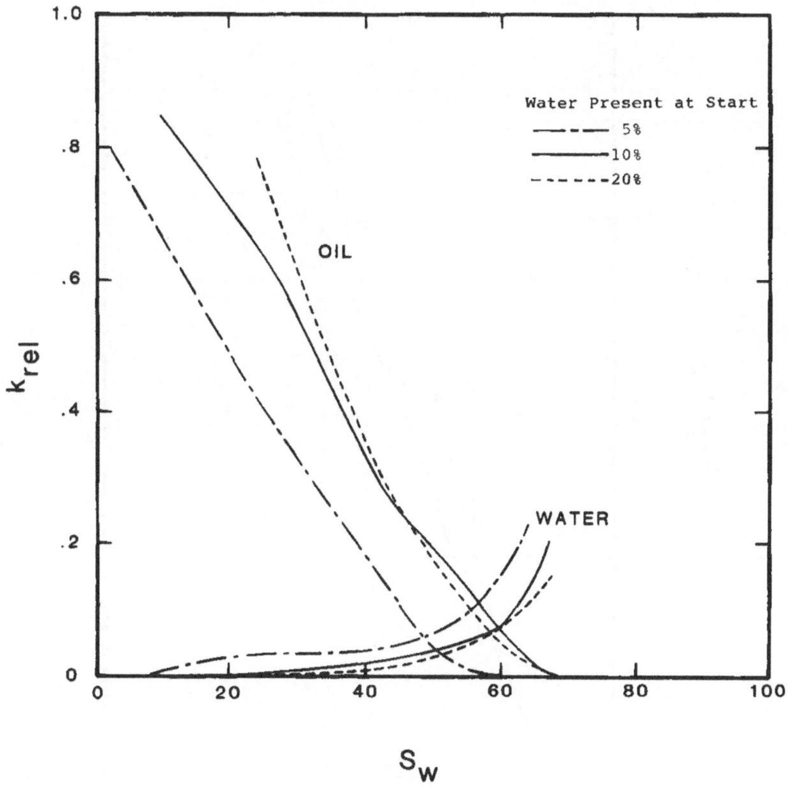

FIGURE 46. Effect of original water saturation on relative permeability.[14]

laboratory relative permeability experiments with fluids which do not differ greatly in viscosity from the reservoir fluids.

XV. EFFECTS OF INITIAL WETTING-PHASE SATURATION

The amount of initial interstitial water affects the oil-water relative permeability values. Caudle et al.[14] investigated this relationship. Figure 46 shows the effect of varying the amount of initial water saturation on water and oil relative permeability. It can be seen that not only the starting points, but also the shape of the relative permeability curves vary with the amount of initial interstitial water.[105]

Sarem[172] found that the presence of initial water saturation tended to shift water-oil relative permeability ratio curves toward the region of lower oil saturation. The difference in the residual oil saturation caused by this shift was reported to be about half the difference in initial water saturation. Thus, a lower residual oil saturation is obtained at higher values of initial water saturation.

Henderson et al.[35,165] noted that the maximum effect of initial water saturation on the relative permeability curve was a shift of the entire curve laterally approximately 4% along the saturation axis, in a direction which increased the oil saturation for a given pair of relative permeability values. Craig indicated that up to 20% initial connate water saturation in oil-wet cores had no effect on oil-water relative permeabilities. However, a definite effect was observed in water-wet cores.

It is suggested that, except for special studies, the amount of water present at the start of a relative permeability determination should be the irreducible water saturation of the sample.

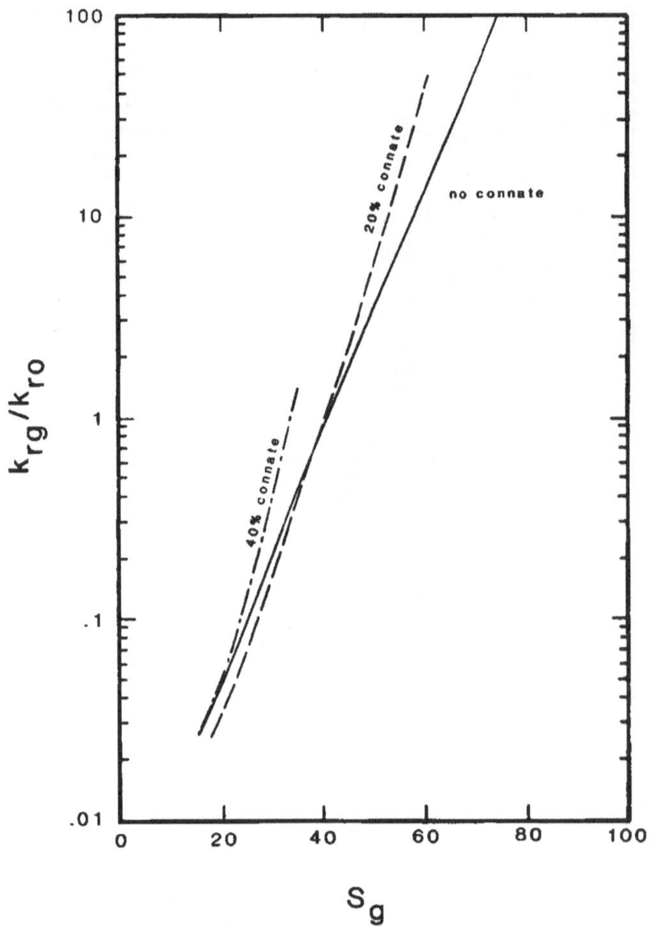

FIGURE 47. Effect of connate water on relative permeability ratio.[174]

XVI. EFFECTS OF AN IMMOBILE THIRD PHASE

Many hydrocarbon reservoirs have only two mobile fluid phases. The mobile phases may be gas and oil in the upper portion of the reservoir and water and oil in the lower portion. Thus, two-phase relative permeabilities are sufficient to characterize fluid flow behavior in these reservoirs.

Some investigators suggest that the immobile water saturation may be regarded as part of the rock, and gas and oil saturations may be given in terms of the hydrocarbon pore space. Owens et al.[155,173] tested several native-state and cleaned cores, both water-wet and oil-wet, and found that an immobile connate water saturation had no measurable influence on the gas-oil relative permeability ratio in the majority of the cases that were studied. Calhoun[174] concluded that low water saturations did not appreciably affect the permeability ratio, simply because the water occupies space which does not contribute to the flow capacity of the rock. Figure 47 shows the effect of connate water saturation on gas-oil permeability ratio. Stewart et al.[175] have also shown that in a limestone with intergranular porosity, the effect of interstitial water on external gas or solution gas drive gas-oil relative permeability ratio is negligible.

Leas et al.[121] reported a close agreement between the gas-oil relative permeability of a system at various values of interstitial water saturation. This agreement was best in the

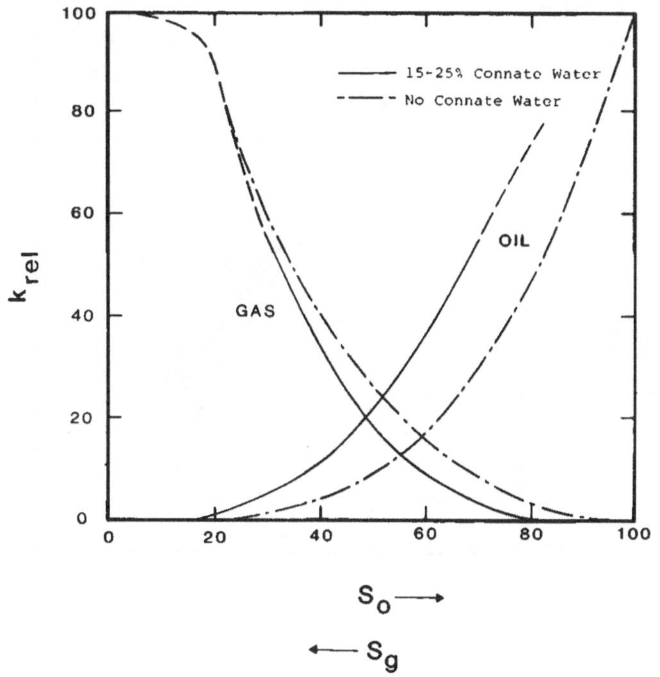

FIGURE 48. Effect of the presence of connate water on relative permeabilities.[4]

equilibrium gas saturation region. They concluded that the gas relative permeability is dependent on total liquid saturation. Other investigators have suggested that even though the immobile connate water does not appreciably affect the relative permeability ratio, the amount and distribution of the interstitial water may influence the relative permeability curve. Dunlap,[118] Leverett,[4] Caudle et al.,[14] and McCaffery[59] have indicated a dependency on connate water saturation. Figure 48 compares the permeability-saturation curves for oil and gas at 15 to 25% connate water with the corresponding curves without connate water.

Kyte et al.[176] studied a wide range of core materials and fluid properties that could influence residual saturation, to determine the mechanism of oil displacement by water in a partially gas-saturated porous system. They found that the initial gas saturation is related to the trapped gas saturation, which plays a beneficial role in reducing residual oil saturation. Mattax and Clotheir[177] concluded that the trapped gas saturation could improve oil-water relative permeability values in consolidated water-wet sandstones. (See Figure 49.)

Holmgren and Morse[178] attributed the oil recovery improvement of a sample in the presence of residual gas to one or more of the following factors:

1. The changes in physical characteristics of oil.
2. The selective plugging action of the gas as indicated by Kyte.
3. Inclusion of mist in the free gas phase.
4. The additional sweeping or driving action of the free gas as indicated by Leverett.[4,8]

Holmgren and Morse concluded that the changes in physical characteristics of oil, within the pressure range used for their experimental work, were not sufficient to account for the differences in the residual oil saturation which were noted. They further stated that a change in displacement mechanism was the most important cause of the oil recovery improvement.

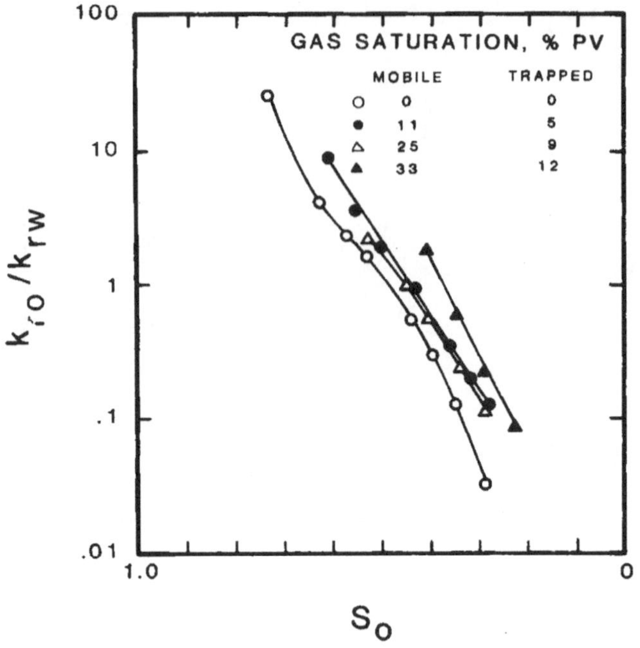

FIGURE 49. Water-oil relative permeability ratio improvement due to trapped gas.[177]

Schneider and Owens[84] investigated the effect of trapped gas saturation in sandstone and carbonate rocks and concluded that the trapped gas affected water relative permeability more than oil relative permeability in oil-wet rocks. These effects are illustrated in Figures 50 and 51. They also concluded that the trapped gas saturation lowered the maximum value of oil relative permeability. Water relative permeability was also lowered as a result of an increase in trapped gas saturation. These effects are illustrated by Figure 52.

XVII. EFFECTS OF OTHER FACTORS

The effects of displacement pressure, pressure gradient, and flow rate on the shape of relative permeability curves have long been a controversial subject in petroleum-related literature. Some authors believe that the effect of displacment pressure and pressure gradient may be due to the changes imposed on viscosity, interfacial tension, and other fluid or rock properties. Others believe that the changes in relative permeability, which appear to result from changes in displacment pressure and pressure gradient, are actually due primarily to an "end effect" developed during laboratory tests.

End effect or boundary effect refers to a discontinuity in the capillary properties of a system at the time of relative permeability measurement. In a stratum of permeable rock, the capillary forces act uniformly in all directions, and thus negate each other. In a laboratory sample, however, there is a saturation discontinuity at the end of a sample. When the flowing phases are discharged into an open region under atmospheric pressure, a net capillary force persists in the sample; this force tends to prevent the wetting phase from leaving the sample. The accumulation of the wetting phase at the outflow face of the sample creates a saturation gradient along the sample which disturbs the relative permeability measurements. For example, a large difference in saturation at the displacement front causes a large capillary pressure gradient, which in turn causes the water to advance ahead of the flood front and to reduce the capillary pressure gradient in the measured region. The advancing water cannot be produced when it first reaches the outflow face of a core, because the pressure in the

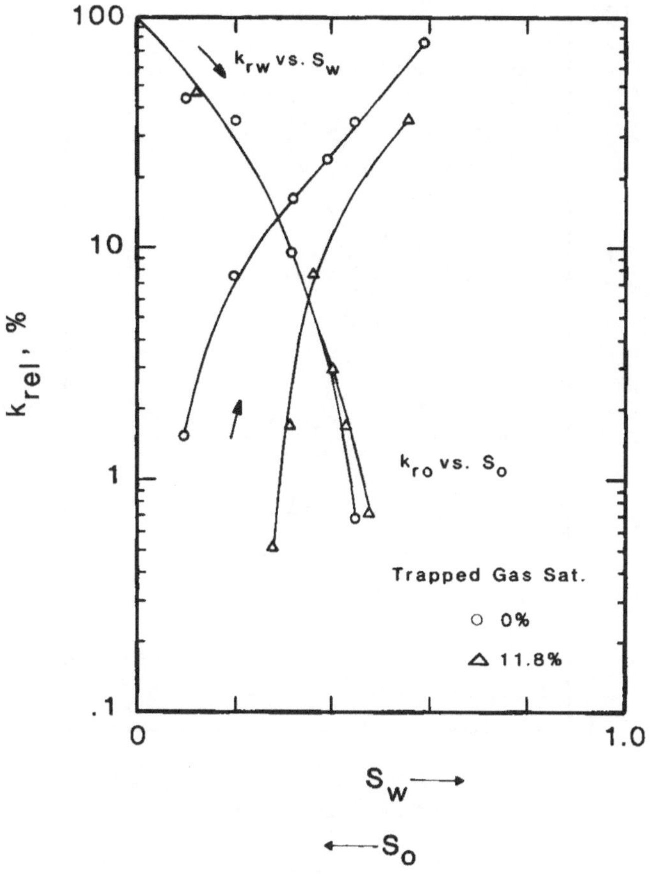

FIGURE 50. Effect of trapped gas saturation (oil wet Grayburg carbonate).[84]

water just inside the core is lower than the pressure in the oil-filled space around the outflow face. This difference in pressure is equal to the capillary pressure for the existing saturation at the outflow face. Therefore, water accumulates at the outflow end of the core, causing a reduction in the capillary pressure. The water will not be produced until the capillary pressure is overcome and the residual oil saturation (at the outflow face of the core) is reached. The calculation of relative permeability based on the average saturation of the sample produces erroneous results in this case, since the relative permeability varies throughout the core due to the saturation gradient created by the wetting phase accumulation at the outflow face of the core.

Owens et al.,[155] Sandberg et al.,[140] Kyte and Rappoport,[156] and Perkins[149] believe that the most convenient way of minimizing the boundary effect is the adjustment of capillary forces to insignificant values, as compared to the viscous forces. This is usually done by a flow rate adjustment. However, the adjusted rate must be low enough so the inertial forces do not disturb the laboratory measurement. It is suggested that the higher flow rate also increases the fluid dispersion at the inflow end of the sample, so that fluid mixing is enhanced. An equation has been developed[157] to predict the extent that a core can be disturbed by boundary effect, at a given rate. Another convenient way of minimizing the boundary effect at the outflow end of a core is to use a more viscous oil in a longer core.[156]

Leverett et al.[4,8] reported, then refuted, the influence of flow rate upon relative permeability. They eventually attributed the observed deviations in their results to an end effect,

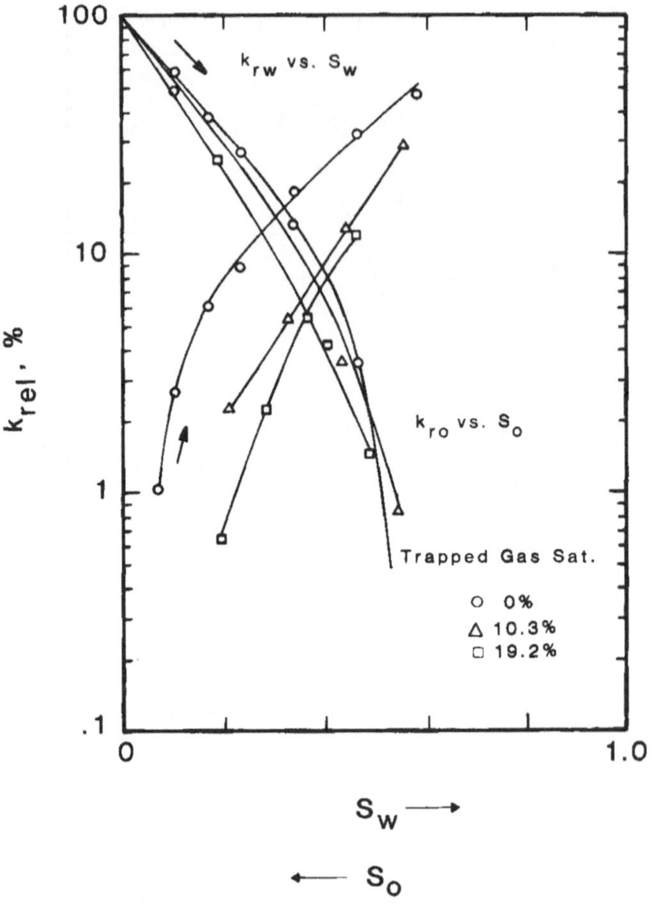

FIGURE 51. Effect of trapped gas saturation (oil wet Tensleep sandstone).[84]

such as that previously described by Hassler.[1] Crowell et al.[30] found that a 50-fold variation of injection rate, within the limits of viscous flow of water and gas, had no effect on residual gas saturation of an Arizona sandstone. Geffen et al.[142] also concluded that, at reasonable flow rates, the effect of waterflooding rate on the efficiency of gas displacement was negligible. Henderson and Yuster[35] and Morse et al.[158] found that relative permeability was rate-dependent in all gas-liquid systems that were studied. Wyckoff and Botset[3] also found that the gas and liquid relative permeabilities were rate-dependent when the two phases were allowed to flow through the core under the same pressure gradient.

Caudle et al.[14] found that relative permeability decreased with increase in flow rate when one of the flowing phases was a gas. Labastie et al.,[159] however, investigated the effect of flow rate in a water-wet sandstone and oil-wet carbonate cores and concluded that relative permeabilities were independent of flow rate except near residual oil saturation. Sandberg et al.,[140] Richardson et al.,[157] Osoba et al.,[13] and Leas et al.[121] found that drainage relative permeability is independent of the flow rate as long as a saturation gradient is not introduced in the core by the inertial forces. Pirson[102] concluded that relative permeability is not rate-sensitive in drainage processes. Ehrlich and Crane[148] examined the effect of flow rate variation on steady-state relative permeability and concluded that both imbibition and drainage relative permeability were independent of flow rate.

Handy and Datta[162] found that the imbibition relative permeability values were dependent on the imbibition procedures; that is, the relative permeability values under free imbibition

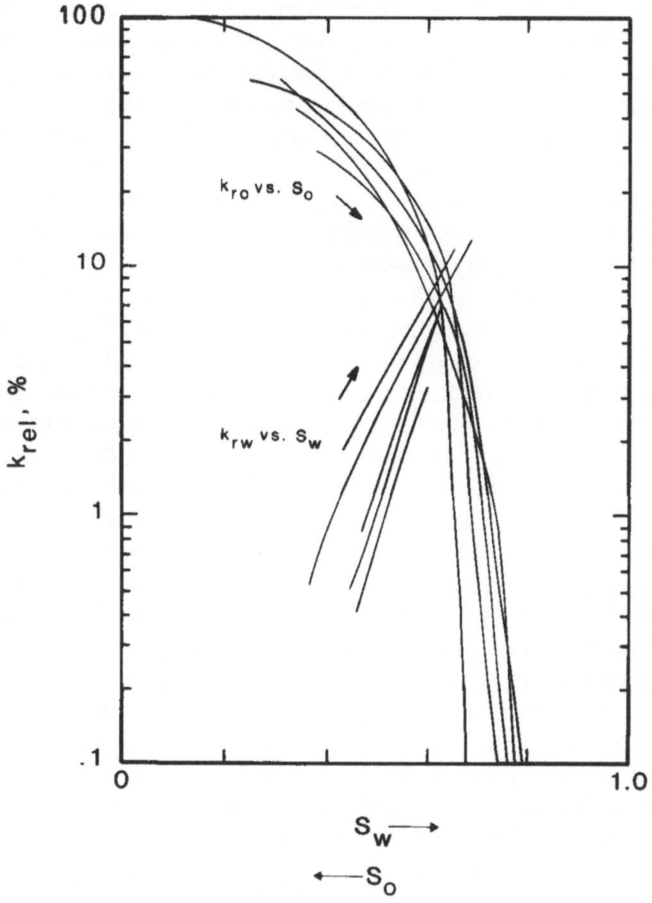

FIGURE 52. Effect of trapped gas saturation (water wet Tensleep sandstone).[84]

were larger than those under a controlled process. The difference between free and controlled imbibition was found to be smaller for more permeable samples. Perkins[149] found that the residual oil saturation after flooding was independent of the flooding rate and concluded that capillary forces controlled the microscopic fluid distribution in the core. Moore and Slobod[67] reported that waterflood recovery from a water-wet core was practically independent of flooding rate. However, they observed that a significant recovery increase may be obtained at extremely high rates. Huppler[163] stated that the waterflood recovery from cores with significant heterogeneity was sensitive to flooding rate. Lefebvre du Prey[154] concluded that the relative permeability was a function of velocity (v), through the ratio ($\sigma\cos\theta/\mu v$), when the viscous forces predominate.

Wyckoff and Botset,[3] Leverett,[4] and Henderson et al.[35,165] studied the possible effects of displacement pressure and pressure gradient on water-oil relative permeability. They concluded that the water and oil relative permeability values were slightly influenced by these factors. Muskat[134] and Krutter and Day,[166] however, reported that the gas and oil relative permeability values of a consolidated sandstone were not affected by changes in differential pressure. McCaffery[179] indicated that the drainage relative permeability values were not influenced by the flow rates which result from a pressure gradient in the range of 1.0 to 5.0 psi across a 12 in. core. Delclaud[160] also concluded that relative permeability is independent of displacement pressure. Pirson,[102] however, suggested that the relative permeability in an imbibition cycle is sensitive to pressure gradient.

Krutter and Day[166] found that ultimate recovery increases with increasing pressure gradient, although the ratio of increased recovery to increased pressure gradient decreases in the region of high pressure gradients. Brownell and Katz[168] reported that an increase in pressure gradient decreased the residual saturation toward zero in the systems that were investigated. Geffen et al.[142] also confirmed that residual gas saturation was a function of pressure gradient. Stegemeier and Jensen[37] believed that the residual wetting phase in a drainage process was held in pendular rings interconnected with only thin wetting-phase layers. They concluded that this residual wetting phase was trapped by capillary forces and that a higher pressure gradient might overcome the capillary pressure and reduce the residual wetting-phase saturation.

Stewart et al.[119] observed that the rate of pressure decline in a nonuniform limestone might influence the gas-oil relative permeability ratio when the solution gas displacement technique for relative permeability measurement was employed. Wall and Khurana[169] found the gas saturation developed in a sand pack, at a given rate of pressure decline, was a function of the mean particle size and probably a function of permeability. They found that a finer grain sand pack gave rise to higher gas saturations in the solution gas displacement technique.

Crowell et al.[30] studied the effect of core dimensions on laboratory measurement of relative permeability. They found that the residual gas saturation in water-gas systems was almost independent of length of the core, within limits of the laboratory-scale models used. They also examined cylindrical and rectangular samples, and observed that a 100-fold change in the ratio of core length to core cross-sectional area of Berea and Boise sandstones did not alter the residual gas saturation of the samples. Moore and Slobod[67] also found that fluid recovery from water-wet cores was not affected by the sample length. Perkins[149] and McCaffery[179] recommended the use of longer cores, to reduce influence of the end effect.

Rose[170] studied the effect of gas expansion, created by the pressure gradient along the sample, on gas-liquid flow characteristics. He concluded that a necessary condition for correct steady-state measurements of liquid-gas relative permeability was the establishment of a uniform fluid saturation distribution in the core. Osoba et al.[13] found that gas expansion affected gas and oil relative permeability values in tests conducted at near-atmospheric pressure. Richardson et al.,[157] however, found that the effect of gas expansion on gas and oil relative permeability values was insignificant at the low pressures which were employed in their study.

In the laboratory gas displacement method of relative permeability measurement, a "stabilized zone" tends to form when the wetting liquid saturation is sufficiently high to permit its readjustment faster than the imposed displacement by the external drive. The relative permeability values obtained prior to passage of the stabilized zone are not valid. Therefore, it is advantageous to reduce the range of saturation influenced by the stabilized zone, to obtain valid measurements over as wide a saturation range as possible.

It can be shown from the Buckley-Leverett equation that the saturation at which the stabilized zone passes out of a system is inversely related to the viscosity of the displaced liquid. This relationship is based on an assumption that a true stabilized zone forms in laboratory gas drives on short cores. It can also be shown that the length of the stabilized zone is inversely related to the injection rate or differential pressure. It has been suggested that the stabilized zone will be sufficiently small if the pressure differential is of such a magnitude that a volume of gas approximately equal to one half the pore volume of the sample would be produced in less than 60 sec. This flow rate insures that the portion of the core in which the capillary effects predominate will be a negligibly small fraction of the total pore space. Loomis and Crowell[171] showed experimentally that the influence of the stabilized zone fluid flow is much less marked with relatively viscous oil as the displaced phase.

Botset[21] investigated the effect of saturation pressure on gas-oil permeability values and concluded that the saturation pressure had negligible effect on laboratory relative permeability measurement. Stewart studied the effect of gas supersaturation on laboratory solution gas displacement relative permeability measurements. He indicated that even though very little supersaturation exists under most field conditions, the effect may be significant for laboratory tests conducted at high flow rates. He found that the gas-oil relative permeability ratio was generally independent of the degree of supersaturation in rock with intergranular porosity.

The influence of dispersion on relative permeability was studied by Chilingarian et al.[99] They concluded that an increase in degree of dispersion increased the relative permeability of the porous medium to both the continuous and discontinuous phases. They also concluded that the degree of dispersion increased with decreasing interfacial tension and increasing time of coalescence of dispersed-phase droplets.

REFERENCES

1. **Hassler, G. L., Rice, R. R., and Leeman, E. H.,** Investigations of recovery on the oil from sandstones by gas-drive, *Trans. AIME,* 118, 116, 1936.
2. **Muskat, M. and Meres, M. W.,** *Physics,* 7, 346, 1936.
3. **Wyckoff, R. D. and Botset, H. G.,** Flow of gas liquid mixtures through sands, *Physics,* 7, 325, 1936.
4. **Leverett, M. C.,** Flow of oil-water mixtures through unconsolidated sands, *Trans. AIME,* 132, 149, 1939.
5. **Nowak, T. J. and Krueger, R. P.,** The effect of mud filtrates and mud particles upon the permeability of cores, Proceedings of the Spring API Meeting, Los Angeles, 1955.
6. **Yuster, S. T.,** Theoretical Consideration of Multiphase Flow in Idealized Capillary System, Proceedings of the Third World Petroleum Congress, Hague, Netherlands, 1951, (2) 437.
7. **Odeh, A. S.,** Relative Permeability Studies, Masters thesis, University of California, Los Angeles, 1953.
8. **Leverett, M. C. and Lewis, W. B.,** Steady flow of gas-oil-water mixtures through unconsolidated sands, *Trans. AIME,* 142, 107, 1941.
9. **Sarem, A. M.,** Three-phase relative permeability measurements by unsteady-state methods, *Soc. Pet. Eng. J.,* 9, 199, 1966.
10. **Saraf, D. N. and Fatt, I.,** Three-phase relative permeability measurement using a N.M.R. technique for estimating fluid saturation, *Soc. Pet. Eng. J.,* 9, 235, 1967.
11. **Owens, W. W. and Archer, D. L.,** The effect of rock wettability on oil-water relative permeability relationships, *Trans. AIME,* 251, 873, 1971.
12. **Geffen, T. M., Owens, W. W., Parrish, D. R., and Morse, R. A.,** Experimental investigation of factors affecting laboratory relative permeability measurements, *Trans. AIME,* 192, 99, 1951.
13. **Osoba, J. S., Richardson, J. G., Kerver, J. K., Hafford, J. A., and Blair, P. M.,** Laboratory measurements of relative permeability, *Trans. AIME,* 192, 47, 1951.
14. **Caudle, B. H., Slobod, R. L., and Brownscombe, E. R.,** Further developments in the laboratory determination of relative permeability, *Trans. AIME,* 192, 145, 1951.
15. **Snell, R. W.,** Measurement of gas-phase saturation in porous media, *J. Inst. Pet.* 45, (428), 1959.
16. **Emmett, W. R., Beaver, K. W., and McCaleb, J. A.,** Little Buffalo basin Tensleep heterogeneity and its influence on drilling and secondary recovery, *J. Pet. Technol.,* 2, 161, 1971.
17. **Donaldson, E. C. and Dean, G. W.,** Two- and Three-Phase Relative Permeability Studies, report # 6826, U.S. Department of the Interior, Bureau of Mines, Bartlesville, Okla., 1966.
18. **Arps, J. J. and Roberts, T. G.,** The effect of the relative permeability ratio, the oil gravity, and the solution gas-oil ratio on the primary recovery from depletion type reservoir, *Trans. AIME,* 204, 120, 1955.
19. **Bulnes, A. C. and Fittings, R. U.,** An introductory discussion of reservoir performance of limestone formations, *Trans. AIME,* 160, 179, 1945.
20. **Stone, H. L.,** Probability model for estimating three-phase relative permeability, *Trans. AIME,* 249, 214, 1970.
21. **Botset, H. G.,** Flow of gas liquid mixtures through consolidated sand, *Trans. AIME,* 136, 91, 1940.
22. **Naar, J., Wygal, R. J., and Henderson, J. H.,** Imbibition relative permeability in unconsolidated porous media, *Trans. AIME,* 225, 13, 1962.
23. **Nind, T. E. W., Ed.,** *Principles of Oil Production,* McGraw Hill, New York, 1964.
24. **Corey, A. T. and Rathjens, C. H.,** Effect of stratification on relative permeability, *Trans. AIME,* 207, (358), 69, 1956.
25. **Huppler, J. D.,** Numerical investigation of the effects of core heterogeneities on waterflood relative permeability, *Soc. Pet. Eng. J.,* 10, 381, 1970.

26. **Johnson, C. E., Jr. and Sweeney, S. A.,** Quantitative measurement of flow heterogeneity in laboratory core samples and its effect on fluid flow characteristics, paper SPE 3610 presented at the SPE 46th Annual Meeting, New Orleans, October 3, 1971.
27. **Muskat, M., Wyckoff, R. D., Botset, H. G., and Meres, M. W.,** Flow of gas-liquid mixtures through sands, *Trans. AIME*, 123, 69, 1937.
28. **Morgan, T. J. and Gordon, D. T.,** Influence of pore geometry on water-oil relative permeability, *J. Pet. Technol.*, 1199, 407, 1970.
29. **Gorring, R. L.,** Multiphase Flow of Immiscible Fluids in Porous Media, Ph.D. thesis, University of Michigan, Ann Arbor, 1962.
30. **Crowell, D. C., Dean, G. W., and Loomis, A. G.,** Efficiency of gas displacement from a water drive reservoir, *U.S. Bureau Mines*, 6735, 30, 1966.
31. **Fatt, I.,** Network model of porous media, dynamic properties of networks with tube radius distribution, *Trans. AIME*, 207, 164, 1956.
32. **Dodd, C. G. and Kiel, O. G.,** Evaluation of Monte Carlo method in studying fluid-fluid displacement and wettability in porous rocks, *J. Phys. Chem.*, 63, 1646, 1959.
33. **Wyllie, M. R. J.,** Interrelationship between wetting and non-wetting phase relative permeability, *Trans. AIME*, 192, 381, 1951.
34. **Pathak, P., Davis, H. T., and Scriven, L. E.,** Dependence of residual nonwetting liquid on pore topology, paper SPE 11016, presented at the SPE 57th Annual Fall Meeting, New Orleans, 1982.
35. **Henderson, J. H. and Yuster, S. T.,** Relative Permeability Study, *World Oil*, 3, 139, 1948.
36. **Land, C. S. and Baptist, O. C.,** Effect of hydration of montmorillonite on the permeability to gas of water-sensitive reservoir rocks, *J. Pet. Technol.*, 10, 1213, 1965.
37. **Stegemeier, G. L. and Jensen, F. W.,** The Relationship of Relative Permeability to Contact Angles, Theory of Fluid Flow in Porous Media Conference, University of Oklahoma, 1959.
38. **Benner, F. C. and Bartell, F. E.,** The effect of polar impurites upon capillary and surface phenomena in petroleum production, *Drill. Prod. Pract.*, 341, 209, 1941.
39. **Salathiel, R. A.,** Oil recovery by surface film drainage in mixed wettability rocks, paper SPE 4104 presented at SPE 47th Annual Meeting, San Antonio, Calif., October 8, 1972.
40. **Reisberg, J. and Doscher, T. M.,** Interfacial phenomena in crude oil-water systems, *Prod. Mon.*, 10, 43, 1956.
41. **Denekas, M. O., Mattax, C. C., and Davis, G. T.,** Effect of crude oil components on rock wettability, *Trans. AIME*, 216, 330, 1959.
42. **Evans, C. R., Rogers, M. A., and Baily, N. J. L.,** *Chem. Geol.*, 8, 147, 1971.
43. **Nutting, P. G.,** Some physical and chemical properties of reservoir rocks bearing on the accumulation and discharge of oil, *Probl. Pet. Geol. AAPG*, 12, 127, 1934.
44. **Leach, R. O., Wagner, O. R., Wood, H. W., and Harpke, C. F.,** A laboratory and field study of wettability adjustment in waterflooding, *J. Pet. Technol.*, 44, 206, 1962.
45. **Mungan, N.,** Interfacial effects in immiscible liquid-liquid displacement in porous media, *Soc. Pet. Eng. J.*, 9, 247, 1966.
46. **Schmid, C.,** The wettability of petroleum rocks and results of experiments to study effects of variations in wettability of core samples, *Erdoel Kohle*, 17(8), 605, 1964.
47. **Kusakov, M. M. et al.,** *Research in Surface Forces*, Deryagin, B. U., Ed., Consultants Bureau, New York, 1963.
48. **Craig, F. F., Jr.,** *The Reservoir Engineering Aspects of Waterflooding Monograph*, Vol. 3, SPE of AIME, Henry L. Doherty Series, Dallas, Tex., 1971.
49. **DeBano, L. F. and Letey, J. L.,** Symposium on Water Repellent Soils, University of Calif., Berkeley, 1969.
50. **Holbrook, O. C. and Bernard, G. G.,** Determination of wettability by dye adsorption, *Trans. AIME*, 213, 261, 1958.
51. **Fatt, I. and Klikoff, W. A., Jr.,** Effect of fractional wettability on multiphase flow through porous media, *J. Pet. Technol.*, 10, 71, 1959.
52. **Brown, R. J. S. and Fatt, I.,** Measurements of fractional wettability of oil field rocks by nuclear magnetic relaxation method, *J. Pet. Technol.*, 11, 262, 1956.
53. **Iwankow, E. N.,** A correlation of interstitial water saturation and heterogeneous wettability, *Prod. Mon.*, 24, 18, 1960.
54. **Gimaludinov, Sh. K.,** The nature of mineral surfaces in oil bearing rocks, *Neft. Gazov. Z.*, 12, 37, 1963.
55. **McGhee, J. W., Crocker, M. E., and Donaldson, E. C.,** Relative Wetting Properties of Crude Oils in Berea Sandstone, Bartlesville Energy Technology Center, Department of Energy, Bartlesville, Okla., BETC/RI-7819, January, 1979.
56. **Wagner, O. R. and Leach, R. O.,** Improving oil displacement efficiency by wettability adjustment, *Trans. AIME*, 216, 65, 1959.

57. **Boneau, D. F. and Clampitt, R. L.,** A Surfactant System for the Oil-Wet Sandstone of the North Burbank Unit, Symposium on Improved Oil Recovery, Tulsa, Arizona, March, 1976.
58. **Morrow, N.,** The effects of surface roughness on contact angle with special reference to petroleum recovery, *J. Can. Pet. Technol.*, 10, 42, 1975.
59. **McCaffery, F. G.,** The Effect of Wettability on Relative Permeability and Imbibition in Porous Media, Ph.D. thesis, University of Calgary, Alberta, Canada, 1973.
60. **Zisman, W. A.,** Contact Angle Wettability and Adhesion Advances in Chemistry, *Am. Chem. Soc.*, 43, 1, 1964.
61. **Melrose, J. C. and Brandner, C. F.,** Role of capillary forces in determination of microscopic displacement efficiency for oil recovery by water flooding, *J. Can. Pet. Technol.*, 10, 54, 1974.
62. **Treiber, L. E., Archer, D. L., and Owens, W. W.,** A laboratory evaluation of the wettability of fifty oil producing reservoirs, *Soc. Pet. Eng. J.*, 12(6), 531, 1972.
63. **Morrow, N. R., Cram, P. J., and McCaffery, F. G.,** Displacement studies in dolomite with wettability control by octanoic acid, *Soc. Pet. Eng. J.*, 13(4), 221, 1973.
64. **Mungan, N.,** Enhanced oil recovery using water as a driving fluid, *World Oil*, 3, 77, 1981.
65. **Amott, E.,** Observations relating to the wettability of porous rock, *Trans. AIME*, 216, 156, 1959.
66. **Raza, S. H., Treiber, L. E., and Archer, D. L.,** Wettability of reservoir rocks and its evaluation, *Prod. Mon.*, 32, 156, 1968.
67. **Moore, T. F. and Slobod, R. L.,** The effect of viscosity and capillarity on the displacement of oil by water, *Prod. Mon.*, 8, 20, 1956.
68. **Bobek, J. E., Mattax, C. C., and Denekas, M. O.,** Reservoir rock wettability — its significance and evaluation, *Trans. AIME*, 213, 155, 1958.
69. **Killens, C. R., Nielsen, R. F., and Calhoun, J. C.,** Capillary Desaturation and Imbibition in Porous Rock Mineral Industries, Experimental Station Bulletin #62, Penn State University, University Park, 1953, 55.
70. **Richardson, S. G.,** Flow Through Porous Media, *Handbook of Fluid Dynamics Section 16*, McGraw-Hill, New York, 1961.
71. **Donaldson, E. C., Thomas, R. D., and Lorenz, P. B.,** Wettability determination and its effect on recovery efficiency, *Soc. Pet. Eng. J.*, 3, 13, 1969.
72. **Mungan, N.,** Role of wettability and interfacial tension in waterflooding, *Soc. Pet. Eng. J.*, 6, 115, 1964.
73. **Emery, L. W., Mungan, N., and Nicholson, R. W.,** Caustic slug injection in the Singleton field, *J. Pet. Technol.*, 12, 1569, 1970.
74. **Kyte, J. R., Nuamann, V. O., and Mattax, C. C.,** Effect of reservoir environment on water-oil displacement, *J. Pet. Technol.*, 6, 579, 1961.
75. **Gatenby, W. A. and Marsden, S. S.,** Some wettability characteristics of synthetic porous media, *Prod. Mon.*, 22, 5, 1957.
76. **Johansen, R. T. and Dunning, H. N.,** Relative Wetting Tendencies of Crude Oils by Capillarimetric Method, U.S. Bureau of Mines, 1961, 5752.
77. **Adams, N. K.,** *The Physics and Chemistry of Surfaces*, 3rd ed., Oxford Univ. Press, London, 1959, 192.
78. **Slobod, R. L. and Blum, H. A.,** Method for determining wettability of reservoir rocks, *Trans. AIME*, 195, 1, 1952.
79. **Lorenz, P. B., Donaldson, E. C., and Thomas, R. D.,** Use of Centrifuge Measurements of Wettability to Predict Oil Recovery, U.S. Bureau of Mines, 1974, 7873.
80. **Reznik, A. A., Fulton, P. F., and Colbeck, S. C., Jr.,** A mathematical imbibition model with fractional-wettability characteristics, *Prod. Mon.*, 31(9), 22, 1967.
81. **Coley, F. H., Marsden, S. S., and Calhoun, J. C., Jr.,** Study of the effect of wettability on the behavior of fluids in synthetic porous media, *Prod. Mon.*, 20(8), 29, 1956.
82. **Keelan, D. K.,** A critical review of core analysis techniques, *J. Can. Pet. Technol.*, 6, 42, 1972.
83. **Poettmann, F. H., Caudle, B. H., Craig, F. F., Jr., Crawford, P. B., Bond, D. C., Farouq Ali, S. M., Holott, C. R., Johansen, R. T., Mungan, N., and Dowd, W. T.,** Secondary and Tertiary Oil Recovery Processes, Interstate Oil Compact Commission, Oklahoma City, Okla., September 1974.
84. **Schneider, F. N. and Owens, W. W.,** Sandstone and carbonate, two- and three-phase relative permeability characteristics, *Soc. Pet. Eng. J.*, 3, 75, 1970.
85. **Scrom, H. M.,** Significance of Water-Oil Relative Permeability Data Calculated from Displacement Tests, Theory of Fluid Flow in Porous Media Conference, University of Oklahoma, 1959, 189.
86. **Amyx, J. W., Bass, D. M., and Whiting, R. L.,** *Petroleum Reservoir Engineering*, McGraw-Hill, New York, 1960.
87. **Colpits, G. P. and Hunter, D. E.,** Laboratory displacement of oil by water under simulated reservoir conditions, *J. Can. Pet. Technol.*, 3(2), 64, 1964.
88. **Haddenhorst, H. G. and Koch, R.,** Effect of temperature and pressure on the separation of solids from petroleum, *Erdoel Kohle*, 2, 12, 1959.

89. **Luks, K. D. and Kohn, J. P.,** The Effect of Methane Under Pressure on the Liquid Solubility of Heavy Hydrocarbon Components, Liquid-Vapor and Solid-Liquid-Vapor Behavior, Progress Report II, API Research Project 135, Notre Dame, Indiana, July, 1971.
90. **Rathmell, J. J., Braun, P. H., and Perkins, T. K.,** Reservoir waterflood residual oil saturation from laboratory tests, *J. Pet. Technol.*, 225, 175, 1973.
91. **Richardson, J. G., Perkins, F. M., Jr., and Osoba, J. S.,** Difference in behavior of fresh and aged east Texas Woodbine cores, *Trans. AIME*, 204, 86, 1955.
92. **Mungan, N.,** Relative permeability measurement using reservoir fluids, *Soc. Pet. Eng. J.*, 12(5), 398, 1972.
93. **Ehrlich, R., Hasiba, H. H., and Raimondi, P.,** Alkaline waterflooding for wettability alteration — evaluation of a potential field application, *J. Pet. Technol.*, 26, 1335, 1974.
94. *Determination of Residual Oil Saturation*, Interstate Oil Compact Commission, Oklahoma City, Okla., 1978.
95. **Jennings, H. H.,** Surface properties of natural and synthetic porous media, *Prod. Mon.*, 21(5), 20, 1957.
96. **Hough, E. W., Rzasa, M. J., and Wood, B. B.,** Interfacial tensions at reservoir pressures and temperatures, apparatus and the water-methane system, *Trans. AIME*, 192, 57, 1951.
97. **Poston, S. W., Ysrael, S., Hossain, A. K., Montgomery, E. F., and Ramey, H. J., Jr.,** The Effect of Temperature on Relative Permeability of Unconsolidated Sands, paper SPE 1897 presented at the SPE 42nd Annual Fall Meeting, Houston, Texas., 1967.
98. **Cuiec, L. E.,** Restoration of the Natural State of Core Samples, paper SPE 5634 presented at the SPE 50th Annual Meeting, Dallas, Tex., 1975.
99. **Chilingarian, G. V., Mannon, R. W., and Rieke, H. H., Eds.,** *Oil and Gas Production From Carbonate Rocks*, Elsevier, Amsterdam, 1972.
100. **Bradley, D. J.,** The Applicability of Wettability Alteration to Naturally Fractured Reservoirs and Imbibition Waterflooding, Masters thesis, University of Tulsa, Oklahoma, 1983.
101. **McCaffery, F. G. and Mungan, N.,** Contact angle and interfacial tension studies of some hydrocarbon water solid systems, *J. Can. Pet. Technol.*, 7, 185, 1970.
102. *Oil Reservoir Engineering*, Pirson, S. J., Ed., McGraw-Hill, New York, 1958, 68.
103. **Lefebvre du Prey, E.,** Deplacements non-miscibles dans les millieux poreux influence des parameters interfaciaux sur les permeabilites relatives, *C.R. IV Coloq. ARTFP Pau*, 1968.
104. **Donaldson, E. C. and Thomas, R. D.,** Microscopic Observations of Oil Displacement in Water-Wet and Oil-Wet Formations, SPE 3555 presented at the 46th SPE Annual Fall Meeting, New Orleans, Oct. 3-6, 1971.
105. **McCaffery, F. G. and Bennion, D. W.,** The effect of wettability on two-phase relative permeabilities, *J. Can. Pet. Technol.*, 10, 42, 1974.
106. **Levine, J. S.,** Displacement experiments in a consolidated porous system, *Trans. AIME*, 201, 57, 1954.
107. **Josendal, V. A., Sandford, B. B., and Wilson, J. W.,** Improved multiphase flow studies employing radioactive tracers, *Trans. AIME*, 195, 65, 1952.
108. **Terwilliger, P. L., Wilsey, L. E., Hall, H. N., Bridges, P. M., and Morse, R. A.,** Experimental and theoretical investigation of gravity drainage performance, *Trans. AIME*, 192, 285, 1951.
109. **Johnson, E. F., Bossler, D. P., and Naumann, V. O.,** Calculation of relative permeability from displacement experiments, *Trans. AIME*, 216, 370, 1959.
110. **Land, C. S.,** Comparison of calculated with experimental imbibition relative permeability, *Trans. AIME*, 251, 419, 1971.
111. **Gardner, G. H. F., Messmer, J. H., and Woodside, W.,** Effective Porosity and Gas Relative Permeability on Liquid Imbibition Cycle, Theory of Fluid Flow in Porous Media Conference, University of Oklahoma, Norman, 1959, 173.
112. **Shelton, J. L. and Schneider, F. M.,** The effect of water injection on miscible flooding methods using hydrocarbons and CO_2, paper SPE 4580 presented at the SPE 48th Annual Meeting, Las Vegas, 1973.
113. **Land, C.,** Calculation of imbibition relative permeability for two- and three-phase flow from rock properties, *Soc. Pet. Eng. J.*, 6, 149, 1968.
114. **Wilson, J. W.,** Determination of Relative Permeability Under Simulated Reservoir Conditions, *AIChEJ*, 2(1), 4, 1956.
115. **Fatt, I. and Barrett, R. E.,** Effect of overburden pressure on relative permeability, *Trans. AIME*, 198, 325, 1953.
116. **Thomas, R. D. and Ward, D. C.,** Effect of overburden pressure and water saturation on gas permeability of tight sandstone cores, *J. Pet. Technol.*, 2, 120, 1972.
117. **Merliss, F. E., Doane, J. D., and Rzasa, M. J.,** Influence of rock and fluid properties and immiscible fluid-flow behavior in porous media, paper 510-G presented at the AIME Annual Meeting, New Orleans, 1955.
118. **Dunlap, E. N.,** Influence of connate water on permeability of sands to oil, *Trans. AIME*, 127, 215, 1938.

119. **Stewart, C. R., Craig, F. F., Jr., and Morse, R. A.,** Determination of limestone performance characteristics by model flow tests, *Trans AIME*, 198, 93, 1953.
120. **Keelan, D. K.,** A practical approach to determination of imbibition gas-water relative permeability, *J. Pet. Technol.*, 4, 199, 1976.
121. **Leas, W. J., Jenks, L. H., and Russell, C. D.,** Relative permeability to gas, *Trans. AIME*, 189, 65, 1950.
122. **Felsenthal, M.,** Correlation of k_g/k_o data with sandstone core characteristics, *Trans. AIME*, 216, 258, 1959.
123. **McCord, D. R.,** Performance predictions incorporating gravity drainage and gas pressure maintenance — LL-370 Area, Bolivar coastal field, *Trans. AIME*, 198, 231, 1953.
124. **Edmondson, T. A.,** Effect of temperature on waterflooding, *Can. J. Pet. Technol.*, 10, 236, 1965.
125. **Poston, S. W., Ysrael, S., Hossain, A. K. M. S., Montgomery, E. F., IV, and Ramey, H. J., Jr.,** The effect of temperature on irreducible water saturation and relative permeability of unconsolidated sands, *Soc. Pet. Eng. J.*, 6, 171, 1970.
126. **Davidson, L. B.,** The effect of temperature on the permeability ratio of different fluid pairs in two-phase systems, *J. Pet. Technol.*, 8, 1037, 1969.
127. **Sinnokrot, A. A., Ramey, H. J., Jr., and Marsden, S. S., Jr.,** Effect of temperature level upon capillary pressure curves, *Soc. Pet. Eng. J.*, 3, 13, 1971.
128. **Weinbrandt, R. M., Ramey, H. J., Jr., and Cassé, F. J.,** The effect of temperature on relative and absolute permeability of sandstones, *Soc. Pet. Eng. J.*, 10, 376, 1975.
129. **Lo, H. Y. and Mungan, N.,** Effect of Temperature on Water-Oil Relative Permeabilities in Oil-Wet and Water-Wet Systems, SPE #4505, Las Vegas, Nev., September 30, 1973.
130. **Sufi, A. S., Ramey, H. J., Jr., and Brigham, W. E.,** Temperature Effects on Relative Permeabilities of Oil-Water Systems, SPE #11701, New Orleans, La., September 26, 1982.
131. **Sufi, A. S., Ramey, H. J., Jr., and Brigham, W. E.,** Temperature Effects on Oil-Water Relative Permeabilities for Unconsolidated Sands, U.S. Department of Energy, Technical Report, 12056-35, December, 1982.
132. **Miller, M. A., and Ramey, H. J., Jr.,** Effect of Temperature on Oil/Water Relative Permeabilities of Unconsolidated and Consolidated Sands, SPE #12116, San Francisco, Calif., October 5, 1983.
133. **Counsil, J. R.,** Steam-Water Relative Permeability, Ph.D. thesis, Stanford Univ., Stanford, Calif., 1979.
134. **Muskat, M.,** *Physical Principles of Oil Production*, McGraw-Hill New York, 1949.
135. **Bardon, C. and Longeron, D.,** Influence of very low interfacial tensions on relative permeability, paper SPE 7609 presented at the SPE 53rd Annual Meeting, Houston, Tex., 1978.
136. **Richardson, J. G.,** Calculation of waterflood recovery from steady-state relative permeability data, *Trans. AIME*, 210, 373, 1957.
137. **Johnson, E. F., Bossler, D. P., and Nauman, V. O.,** Calculation of relative permeability from displacement experiments, *Trans. AIME*, 216, 370, 1959.
138. **Levine, J. S.,** Displacement experiments in a consolidated porous system, *Trans. AIME*, 201, 57, 1954.
139. **Craig, F. F., Jr.,** Errors in calculation of gas injection performance from laboratory data, *J. Pet. Technol.*, 8, 23, 1952.
140. **Sandberg, C. R., Gourney, L. S., Suppel, R. F.,** Effect of fluid flow rate and viscosity on laboratory determination of oil-water relative permeabilities, *Trans. AIME*, 213, 36, 1958.
141. **Donaldson, E. C., Lorenz, P. B., and Thomas, R. D.,** The effect of viscosity and wettability on oil-water relative permeability, paper SPE 1562 presented at the SPE 41st Annual Meeting, Dallas, Oct. 2-5, 1966.
142. **Geffen, T. M., Parrish, D. R., Haynes, G. W., and Morse, R. A.,** Efficiency of gas displacement from porous media by liquid flooding, *Trans. AIME*, 195, 29, 1952.
143. **Krutter, H. and Day, R. J.,** Air-drive experiments on long horizontal consolidated cores, *J. Pet. Technol.*, 12, 1, 1943.
144. **Morse, R. A., Terwilliger, P. K., and Yuster, S. T.,** Relative permeability measurements on small core samples, *Oil Gas J.*, 46, 109, 1947.
145. **Odeh, A. S.,** Effect of viscosity ratio on relative permeability, *Trans. AIME*, 216, 346, 1959.
146. **Baker, P. E.,** Discussion of effect of viscosity ratio on relative permeability, *J. Pet. Technol.*, 219, 65, 1960.
147. **Downie, J. and Crane, F. E.,** Effect of viscosity on relative permeability, *Soc. Pet. Eng. J.*, 6, 59, 1961.
148. **Ehrlich, R. and Crane, F. E.,** A model for two-phase flow in consolidated materials, *Trans. AIME*, 246, 221, 1969.
149. **Perkins, F. M., Jr.,** An investigation of the role of capillary forces in laboratory waterfloods, *J. Pet. Technol.*, 11, 49, 1957.
150. **Pickell, J. J., Swanson, B. F., Hickman, W. B.,** Application of air-mercury and oil-air capillary pressure data in the study of pore structure and fluid distribution, *Soc. Pet. Eng. J.*, 4, 55, 1966.
151. **Warren, J. E. and Calhoun, J. C.,** A study of waterflood efficiency in oil-wet systems, *Trans. AIME*, 204, 22, 1955.

152. **Caro, R. A., Calhoun, J. C., Jr., and Nielsen, R. F.**, Surface active agents increase oil recovery, *Oil Gas J.*, 12, 6, 1952.
153. **Ojeda, E., Preston, F., and Calhoun, J. C., Jr.**, Correlation of residuals following surfactant floods, *Prod. Mon.*, 12, 20, 1953.
154. **Lefebvre du Prey, E. J.**, Factors affecting liquid-liquid relative permeabilities of a consolidated porous medium, *Soc. Pet. Eng. J.*, 2, 39, 1973.
155. **Owens, W. W., Parrish, D. R., and Lamoreaux, W. E.**, An evaluation of a gas drive method for determining relative permeability relationships, *Trans. AIME*, 207, 275, 1956.
156. **Kyte, J. R. and Rapoport, L. A.**, Linear waterflood behavior and end effects in water-wet porous media, *Trans. AIME*, 213, 423, 1958.
157. **Richardson, J. G., Kerver, J. K., Hafford, J. A., and Osoba, J. S.**, Laboratory determinations of relative permeability, *Trans. AIME*, 195, 187, 1952.
158. **Morse, R. A., Terwilliger, P. K., and Yuster, S. T.**, Relative Permeability Measurements on Small Core Samples, *Oil Gas J.*, 46, 109, 1947.
159. **Labastie, A., Guy, M., Delclaud, J. P., and Iffly, R.**, Effect of flow rate and wettability on water-oil relative permeabilities and capillary pressure, paper SPE 9236 presented at the SPE Annual Meeting, Dallas, Tex., Sep. 21-24, 1980.
159a. **McCaffery, F. G.**, The Effect of Wettability on Relative Permeability and Imbibition in Porous Media, Ph.D. thesis, University of Calgary, Alberta, Canada, 1973.
160. **Delclaud, J. P.**, New results on the displacement of a fluid by another in a porous medium, paper SPE 4103 presented at the SPE 47th Annual Meeting, San Antonio, Tex., 1972.
161. **Fetkovitch, M. J.**, The isochronal testing of oil wells, paper SPE 4529 presented at the 48th Annual Fall Meeting of the SPE, Las Vegas, Nevada, 1973.
162. **Handy, L. L. and Datta, P.**, Fluid distributions during immiscible displacements in porous media, *Soc. Pet. Eng. J.*., 10, 261, 1966.
163. **Huppler, J. D.**, Numerical investigation of the effects of core heterogeneities on waterflood relative permeability, *Soc. Pet. Eng. J.*, 12, 381, 1970.
164. **Stewart, C. R. and Owens, W. W.**, A laboratory study of laminar and turbulent flow in heterogeneous porosity limestone, *Trans. AIME*, 213, 121, 1958.
165. **Henderson, J. H. and Moldrum, H.**, Progress report on multiphase-flow studies, *Prod. Mon.*, 4, 12, 1949.
166. **Krutter, A. and Day, R. J.**, Air-drive experiments on long horizontal consolidated cores, *J. Pet. Technol.*, 11, 1, 1943.
167. **Chen, H. K., Counsil, J. R., and Ramey, H. J., Jr.**, Steam-Water Relative Permeability, 1978 Geothermal Resources Council Annual Meeting, Hilo, Hawaii, July 25-27, 1978.
168. **Brownell, L. E. and Katz, D. L.**, Flow of fluids through porous media — single homogeneous fluids, *Chem. Eng. Prog.*, 43(10), 537, 1947.
169. **Wall, C. G. and Khurana, A. K.**, Saturation permeability relationship at low gas saturation, *J. Inst. Pet.*, 57, 261, 1971.
170. **Rose, W. D.**, Fluid distributions characterizing gas-liquid flow, *Trans. AIME*, 192, 372, 1951.
171. **Loomis, A. G. and Crowell, D. C.**, Relative permeability studies. II. Water oil systems, *Prod. Mon.*, 8, 18, 1959.
172. **Sarem, A. M.**, Significance of water-oil relative permeability data calculated from displacement tests, *Proc.*, Theory of Fluid Flow in Porous Media Conference, University of Oklahoma, Norman, 1959, 189.
173. **Owens, W. W., Parrish, D. R., and Lamoreaux, W. E.**, A comparison of field k_g/k_o characteristics and laboratory k_g/k_o test results measured by a new simplified method, paper 518-G presented at the AIME 30th Annul Meeting, New Orleans, 1955.
174. **Calhoun, J. C., Jr.**, *Fundamentals of Reservoir Engineering*, University of Oklahoma Press, Norman, 1947.
175. **Stewart, C. R., Craig, F. F., and Morse, R. A.**, Determination of limestone performance characteristics by model flow tests, *Trans. AIME*, 198, 93, 1953.
176. **Kyte, J. R., Stanclift, J. R., Stephan, S. C., Jr., and Rapoport, L. A.**, Mechanism of waterflooding in the presence of free gas, *Trans. AIME*, 107, 215, 1956.
177. **Mattax, C. C. and Clotheir, A. T.**, Core Analysis of Unconsolidated and Friable Sands, paper SPE 4986 presented at the SPE 49th Annual Meeting, Houston, Tex., 1974.
178. **Holmgren, C. R. and Morse, R. A.**, Effect of free gas saturation on oil recovery by waterflooding, *Trans. AIME*, 192, 135, 1951.
179. **McCaffery, F. G.**, The Effect of Wettability on Relative Permeability and Imbibition in Porous Media, Ph.D. thesis, University of Calgary, Alberta, Canada, 1973.
180. **Gornik, B. and Roebuck, J. F.**, *Formation Evaluation through Extensive Use of Core Analysis*, Core Laboratories, Inc., Dallas, Tx., 1979.

Chapter 4

THREE-PHASE RELATIVE PERMEABILITY

I. INTRODUCTION

Recent innovations in the field of oil recovery have led to a renewed interest in three-phase relative permeability. Three-phase flow occurs when the water saturation is higher than the irreducible level, and oil and gas are also present as mobile phases. Detailed engineering calculations of the performance of reservoirs under recovery methods such as carbon dioxide injection, *in situ* combustion, steam drive, micellar fluid injection, and nitrogen injection frequently require three-phase relative permeability data.

All factors which influence flow in systems containing two mobile phases also apply to three-phase systems. Virtually all oil reservoirs constitute potential three-phase systems, since reservoir rocks invariably contain interstitial water, and naturally occurring oils completely devoid of gas are rare. In fact, a two-phase system of oil and gas may be regarded as a three-phase system in which the water phase is immobile. The number of reservoirs in which oil, gas, and water are simultaneously mobile during primary production is probably small. Nevertheless, three-phase mobility is always possible when a producing interval includes part of the oil-water transitional zone in a reservoir. It is probable, however, that in most cases where oil and free gas are produced with an appreciable water cut, the water is being produced from layers of the reservoir in which relative permeability to water is high and not by true three-phase flow.

In the past, the use of three-phase relative permeability data for conventional reservoir engineering calculations has seldom been necessary. In consequence, considerably less is known about three-phase relative permeability characteristics of rocks than is known for comparable two-phase cases. The realization that detailed engineering calculations of the performance of reservoirs produced by *in situ* combustion processes require three-phase data is quite new. Three-phase relative permeability is useful in the calculation of field performance for reservoirs being produced by simultaneous water and gas drive, and also in analyzing solution gas drive reservoirs which are partially depleted and are being produced by water drive. An increasing interest in three-phase flow phenomena is anticipated.

There are two distinct classes of three-phase relative permeability data: (1) that pertaining to drainage; and (2) that pertaining to imbibition. Drainage refers to the direction of saturation change in which the wetting-phase saturation decreases. Imbibition refers to an increasing wetting-phase saturation. For the relative permeability data to yield correct reservoir predictions, the direction of saturation change in the reservoir must correspond to the direction of saturation change for which the data were derived.

Drainage relative permeability data should be used in the following situations:

1. Enhanced recovery processes involving the injection of dry gas, flue gas, carbon dioxide, and other gases into watered-out reservoirs.
2. Miscible flood processes in which liquified petroleum gas (LPG) is injected into watered-out reservoirs.
3. Production from reservoirs in which the water saturation is greater than the irreducible saturation.

Imbibition relative permeability data should be used under the following conditions:

1. Reservoirs produced by natural water drive.

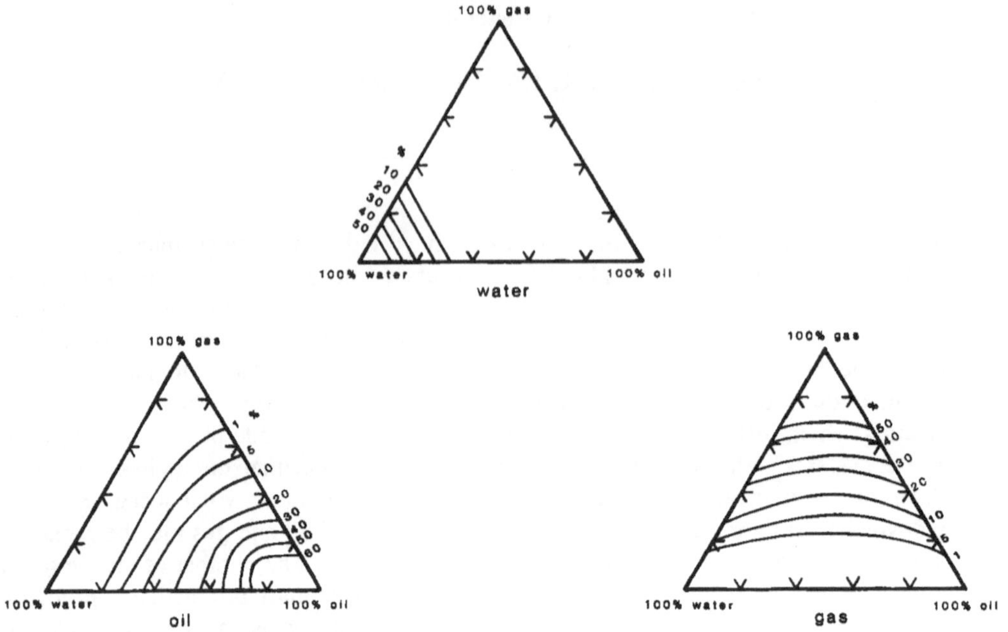

FIGURE 1. Three-phase relative permeability.[1]

2. Reservoirs developed by water flood, as well as by processes where the injected water contains surfactants, polymers, or other additives.
3. Reservoirs developed by recovery processes where water is used to push a slug of chemicals, LPG, etc.

II. DRAINAGE RELATIVE PERMEABILITY

A. Leverett and Lewis

Much of the credit for the classical work in three-phase relative permeability is accorded to Leverett and Lewis[1] who were the first to measure three-phase relative permeability of a water-oil-gas system in an unconsolidated sand. These investigators used a steady-state single-core dynamic method and ignored end effects and hysteresis. Errors from ignoring capillary end effects were probably significant, since low flow rates were used. Ring electrodes were spaced along the length of a sand pack to measure resistivity of the sample and brine saturation was assumed to be directly related to resistivity. Gas saturation was determined from pressure and volume measurements. Oil saturation was obtained by a material balance technique. Leverett and Lewis obtained three separate triangular graphs showing lines of constant relative permeability ("isoperms") to the three phases; these were plotted against the saturations of the three fluids, as shown by Figure 1. They also obtained a plot showing the region of three-phase flow; Figure 2 shows the region where each component comprises at least 5% of the flow stream. As shown by the figure, three-phase flow occurs in a rather confined region.

Relative permeability to water, k_{rw}, was found to be dependent only on water saturation, S_w, and was not affected by the introduction of an additional nonaqueous phase. Relative permeability to gas, k_{rg}, was found to be slightly less than would be expected for the same gas saturation, S_g, in two-phase flow. The k_{rg} isoperms are convex towards the 100% S_g apex of the triangular diagram. As gas becomes one of the two flowing nonwetting phases, when both oil and water are present, the relative permeability to gas decreases as oil saturation

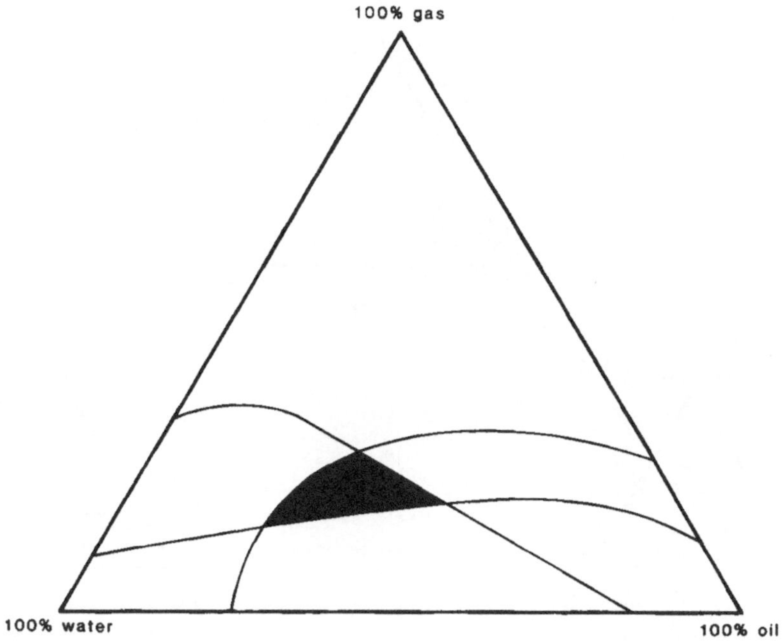

FIGURE 2. Region of three-phase flow.[1]

approaches the water saturation value, becoming a minimum when roughly equal saturations of oil and water are associated with the gas.

The relative permeability to oil is seen to vary in a more complex manner. Starting with a gas saturation of zero, oil relative permeability at constant oil saturation increases as gas saturation increases (except at low oil saturations where k_{rg} remains constant) then decreases to its original value as more gas is introduced, finally falling well below this value when gas saturation is further increased. In a water-wet system, the presence of gas leaves the mode of water flow unchanged, but since the gas tends to occupy the central portions of the intergrain spaces (where the oil is also driven by capillary forces) interference between oil and gas flow is likely. Visual examination under the microscope shows the presence of an oil film (in some cases containing a very small amount of finely divided water) through which oil flows around each gas bubble. It is not clear whether all gas bubbles are connected. However, the gas bubbles are observed to move jerkily, as opposed to the generally smooth flow of water (and of oil when gas bubbles are absent or are stationary). This uneven motion of the gas implies a similar motion of at least part of the oil, which would be expected to move faster than in the absence of gas at the same oil saturation. We see a decrease in k_{ro} at constant S_o as S_g is increased, especially at low S_w.

Also, there is an increase in k_{ro} at constant S_o as S_w is increased at low values of S_w. This effect is evidently due to the shifting of oil into parts of the intergrain space where it may flow more freely. The water introduced tends to occupy the sharply curved parts of the pores, forcing oil into the central space vacated by gas. Since fluid in the sharply curved parts of the pores moves only with difficulty and that in the center moves more readily, the result is an increase in k_{ro}.

Leverett and Lewis pointed out that they found no effect of oil viscosity on the isoperms for various saturations of the three phases.

B. Corey, Rathjens, Henderson, and Wyllie

The results of the work of Corey et al.[2] are shown by Figure 3. These investigators used a calcium chloride brine. Capillary end effects were minimized by using a core with semi-

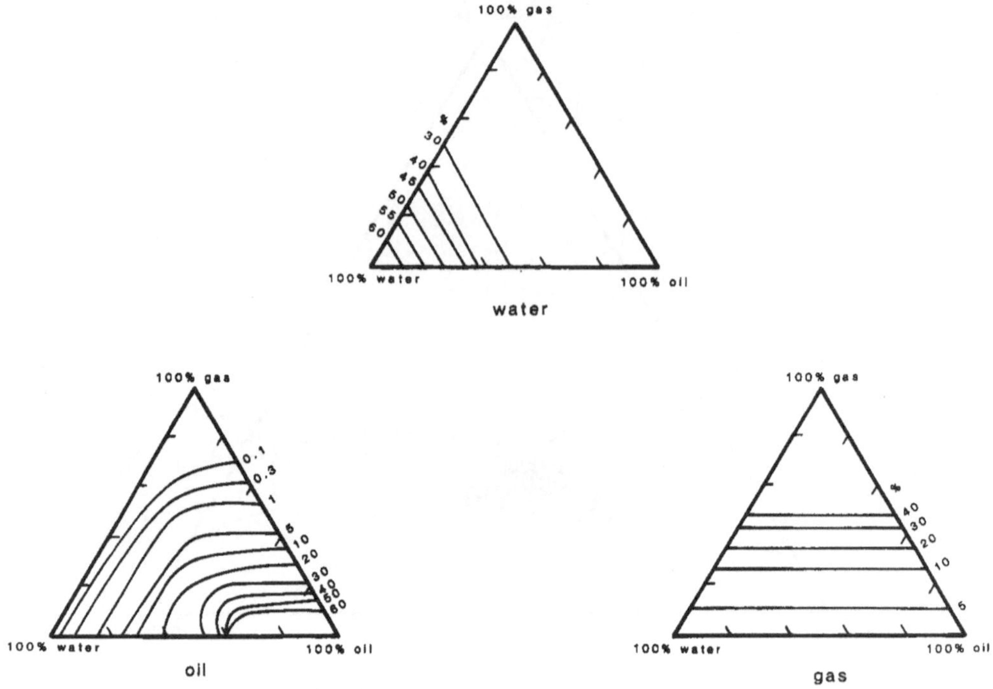

FIGURE 3. Three-phase relative permeability.[2]

permeable membranes mounted at each end. They measured saturations gravimetrically and avoided hysteresis effects by using separate cores for each measurement rather than resaturating the same core. In an initial conclusion, they reported that when the saturations of the wetting phases were equal, the nonwetting phase relative permeability, k_{rn}, was unchanged regardless of whether the nonwetting phase was oil or gas. They used the equivalent liquid permeability as the base value. The oil isoperms of Corey et al. are similar to those obtained by Leverett and Lewis, except that Corey's oil isoperms have a greater curvature. Relative permeability to water was not measured, but was calculated on the assumption that it was a function of water saturation alone and that water permeability in a water-wet system was the same as the oil permeability in an oil-wet system. It should be noted that the data of Corey et al. are for oil drainage in an oil-gas system. They also observed that the behavior of the nonwetting phases was more sensitive to changes in pore geometry than was the behavior of the wetting phase. The increase in k_{ro} (at low S_w) with the increase in S_w (and a corresponding decrease in S_g) is higher in Corey's consolidated sandstone samples than in the unconsolidated samples. This is because of the dependence of k_{ro} on the ratio

$$\frac{\int_{S_w}^{S_L} dS_L/P_c^2}{\int_0^1 dS_L/P_c^2}$$

which is usually higher in consolidated rocks than in unconsolidated rocks.

Corey et al. extended their two-phase relative permeability relationship to three-phase flow on the basis of the following approximation:

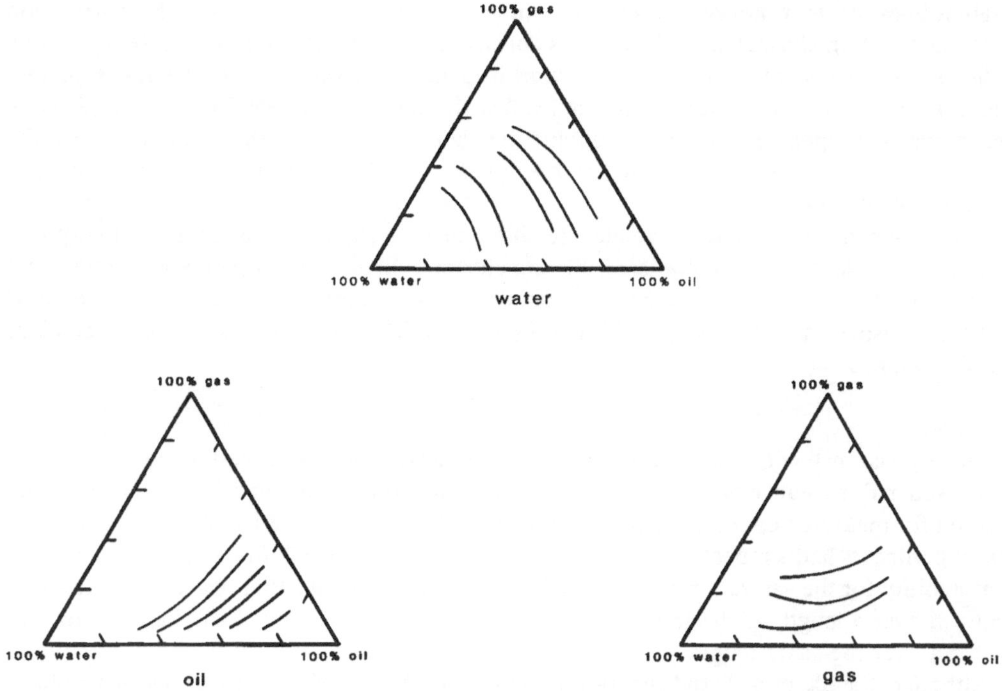

FIGURE 4. Three-phase relative permeability.[3]

$$1/P_c^2 = C \frac{S_{Lr}}{[S_L - (S_{wirr} + S_{or})]} \text{ for } S_L > \frac{S_{Lr}}{(S_{wirr} + S_{or})}$$

$$= 0 \text{ for } S_L \leqq \frac{S_{Lr}}{(S_{wirr} + S_{or})} \tag{1}$$

The drainage oil phase relative permeability in a water-wet system containing gas is given by

$$k_{ro} = \frac{(S_L - S_w)^3}{(1 - S_{Lr})^4} (S_w + S_L - 2S_{Lr}) \tag{2}$$

where S_{Lr} is residual liquid saturation.

As in Leverett's data, the oil isoperms tend to be parallel to the oil isosaturation lines, especially at high S_w. At increasing S_w and constant S_o, the gas which was previously in the system is no longer present. Thus, the rate of increase of k_{ro} with increasing S_w decreases at higher values of S_w. Corey et al. proposed a method to obtain k_{ro} and k_{rw}, based on k_{rg} alone. Incidentally, k_{rg} was found to be a function of S_g and independent of the relative wetting properties of the fluids within the rock.

C. Reid

Using the same method employed by Leverett and Lewis (single-core dynamic technique), Reid[3] obtained the isoperms shown in Figure 4. He eliminated end effects, but hysteresis was ignored. Brine saturation was measured by resistivity, and oil and gas saturations were obtained by gamma ray absorption. His saturation measurements possibly were affected by differential absorption of gamma rays by oil and water. While Leverett and Lewis obtained straight lines for the water phase behavior (showing k_{rw} to be independent of the distribution of the nonwetting phases) and oil isoperms concave toward the 100% S_o apex, Reid's results indicated concave water isoperms, convex oil isoperms, and slightly concave gas isoperms. These results were interpreted as indicating that the relative permeability to each phase is dependent both upon its own saturation and the saturations of the other phases. His results showed a greater oil permeability when three phases were present than with two phases, at a given oil saturation.

Reid made no attempt to correlate the three-phase results with those from two-phase experiments. He placed emphasis on his conclusions for the oil isoperms and noted that Leverett's oil phase data showed a substantial amount of scatter. For this reason, he believed that his oil isoperms were more valid than Leverett's. The work of Rose seems to confirm Reid's findings.

D. Snell

Three-phase behavior in a water-wet unconsolidated sand was investigated by Snell,[4-6] who used radio frequency detection for the determination of S_w and a neutron counting method for measurement of S_g. Oil saturation was obtained by material balance calculation. His experiments had a repeatability within $^1/_2$% for relative permeability values, with a better repeatability for the saturation values. He found that when the wetting phase saturation was uniform over a length of the test sample, the saturations of the other two phases were also uniform over the same length.

Although Caudle et al.[11] did mention hysteresis in their work, the first significant study on the effect of saturation history on three-phase relative permeability was done by Snell. In describing Snell's work, it is convenient to define four types of liquid saturation histories:

1. Imbibition of water with oil saturation increasing (II).
2. Imbibition of water with oil saturation decreasing (ID).
3. Drainage of water with oil saturation increasing (DI).
4. Drainage of water with oil saturation decreasing (DD).

As seen from his results in Figure 5, k_{ro} values were lower for DD than for the other saturation histories. Since, in two-phase flow, drainage caused the wetting phase to lose its mobility at higher saturations, it has been suggested that there is a partial change in wettability from water-wet to oil-wet during DD. When the system was oil-wet, a larger S_o was required for the same k_{ro}, because some of the oil was trapped in the smaller pores. This oil increases S_o, but it is immobile. He further suggested that this change in wettability may be caused by polar compounds in the oil. Snell's results do not show good agreement with those of Leverett and Lewis except in the case of k_{rg}.

Oil and water isoperms reported by Snell are similar to those determined by Reid, but Snell's k_{ro} values are higher than Reid's, especially at low water saturation.

In a later work, Snell reinterpreted the results of four earlier studies done on unconsolidated sands. In these investigations, no hysteresis was found for water isoperms. Oil isoperms showed hysteresis only when kerosene or a kerosene/lubricating oil mixture was used as the oil phase. Nonpolar oil gave no hysteresis. Reinterpretation of the earlier results was possible because Leverett and Lewis indicated possible errors in their saturation measurements. Reid's saturation data might also have been inaccurate because of differential absorption of gamma rays by oil and water. Relative permeability to oil was found to be dependent only on the histories of the liquid phase saturations, although Snell did not rule out dependence on gas phase saturation history. Snell reinterpreted Leverett's data to obtain oil isoperms convex toward the 100% S_o apex. Oil isoperms then followed the same pattern in all four investigations. These results are shown in Figure 6. The curvature of the isoperms of both nonwetting

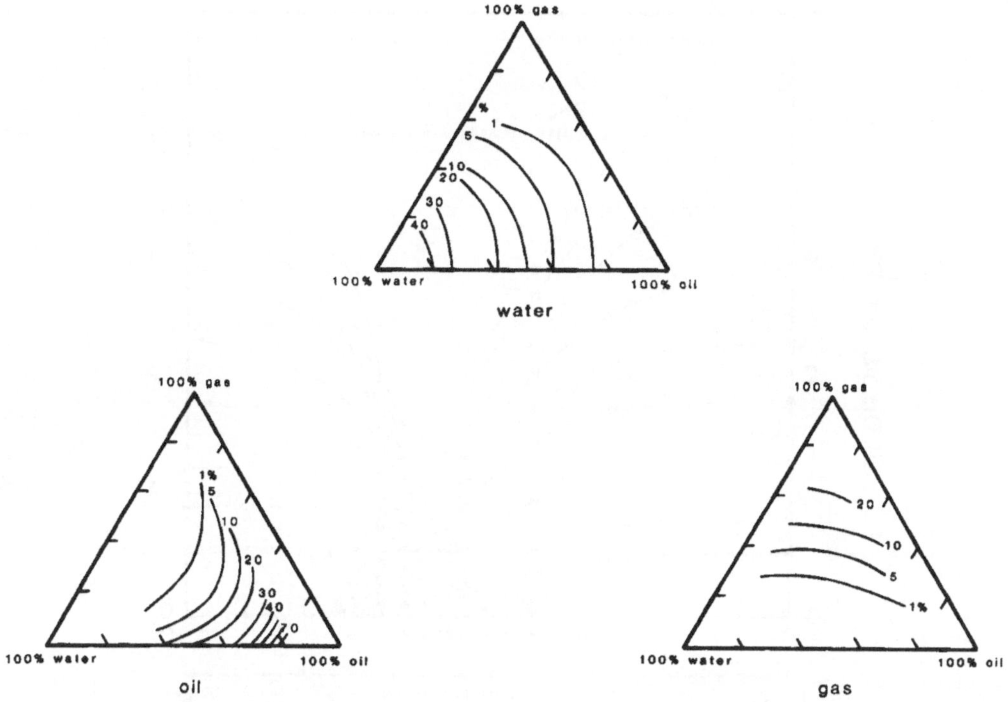

FIGURE 5. Three-phase relative permeability.[5]

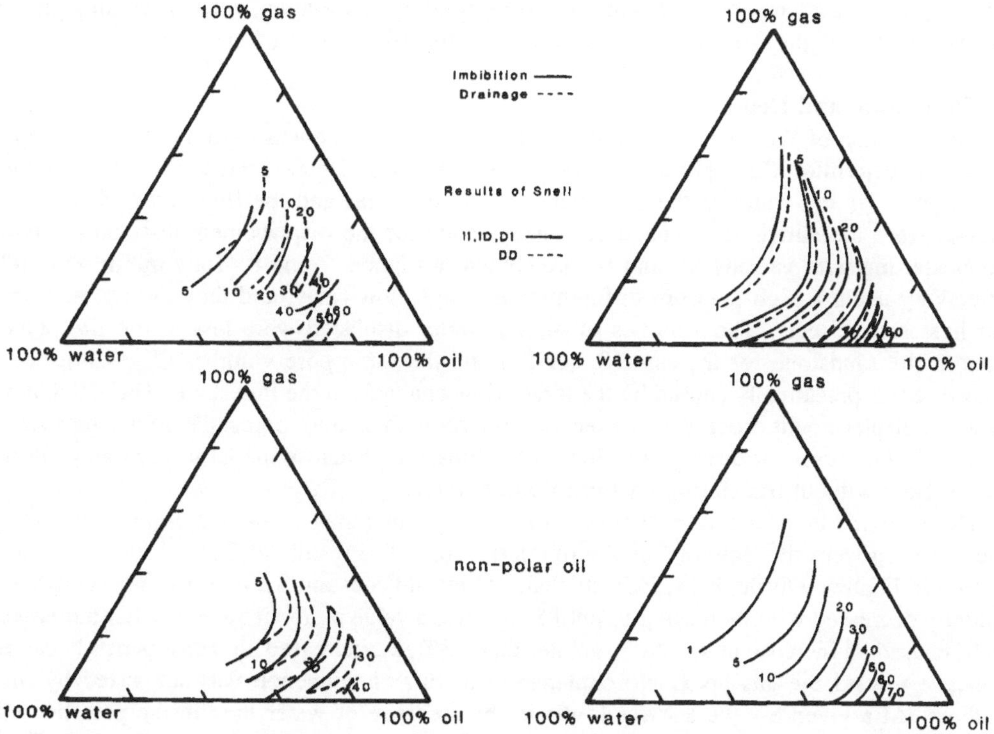

FIGURE 6. Reinterpretation of results by Snell.[6]

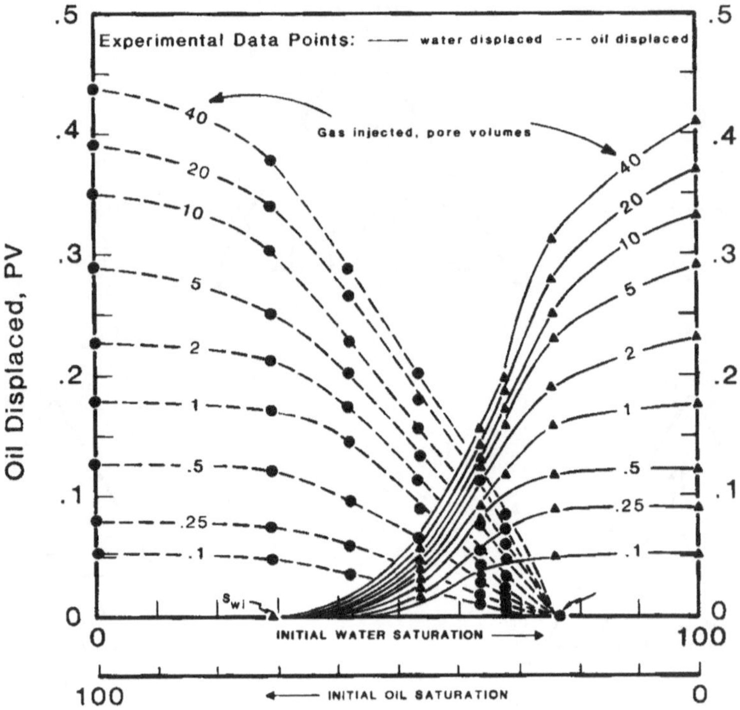

FIGURE 7. Fluid flow experimental data for Berea sandstone.[7]

phases (oil and gas) are convex toward the corresponding phase-apex, whereas wetting phase isoperms are straight lines or are concave toward the 100% apex of the wetting phase.

E. Donaldson and Dean

An extension of Welge's two-phase unsteady-state technique was used by Donaldson and Dean[7] to determine three-phase relative permeabilities of Berea sandstone and Arbuckle limestone. Oil and water in the core were displaced by gas and the flow rates of all three phases were measured simultaneously. Their results for the displacement tests on the two cores starting with various S_{wi} and S_{oi} are shown in Figures 7 and 8. They minimized end effects by using a high pressure differential and high flow rates, and they did not account for hysteresis effects. The volumes of oil and water displaced were less in the limestone than in the sandstone for the same S_o (or S_w) and the same pore volumes of gas injected. This effect is presumably caused by the larger flow channels in the limestone. The efficiency of a gas displacement process is greater for a matrix with smaller pores. There is a narrower range of saturations for three-phase flow in the limestone because the large vugs may allow gas to flow without transfering energy to oil or water.

The isoperms are presented as functions of terminal rather than average saturations, because the former govern the flow of fluids through the core. The results of Donaldson and Dean, shown in Figures 9 through 14, indicate that, at low and constant S_g, k_{rg} for Berea sandstone initially decreased with increasing S_o until S_o reached a value of about 50%. Further increases in S_o caused an increase in k_{rg}. At S_g greater than 13%, k_{rg} increased so the isoperms became concave toward the gas apex. No explanation of this phenomenon was suggested by the authors. At a given S_g, the k_{rg} was lower in the presence of water than in the presence of oil, probably because water adhered more strongly to the rock surface than did the oil. The flow path of gas is more restricted in the presence of water, since gas can displace oil more easily than it can displace water. For the limestone, k_{rg} was always concave toward the gas apex.

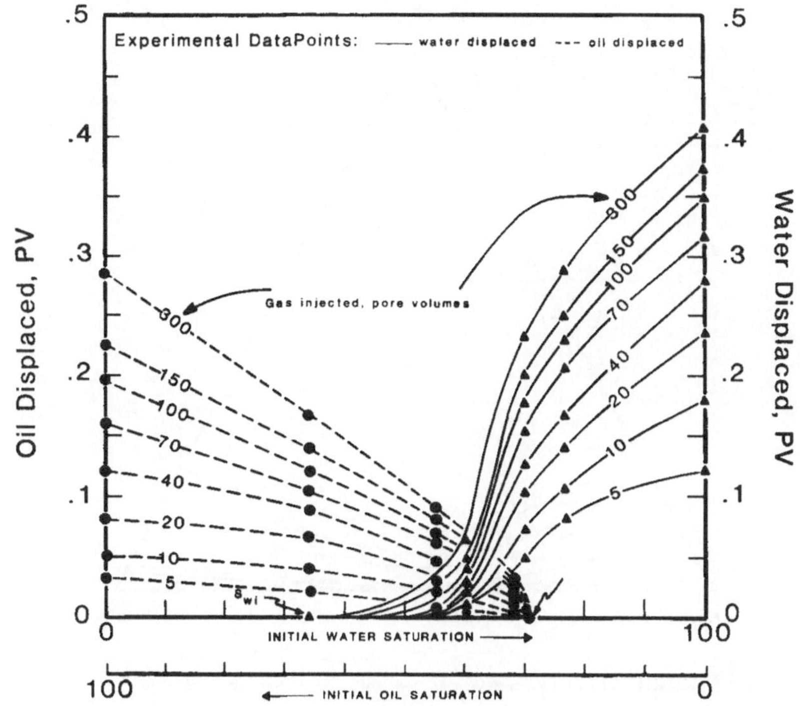

FIGURE 8. Fluid flow experimental data for Arbuckle limestone.[7]

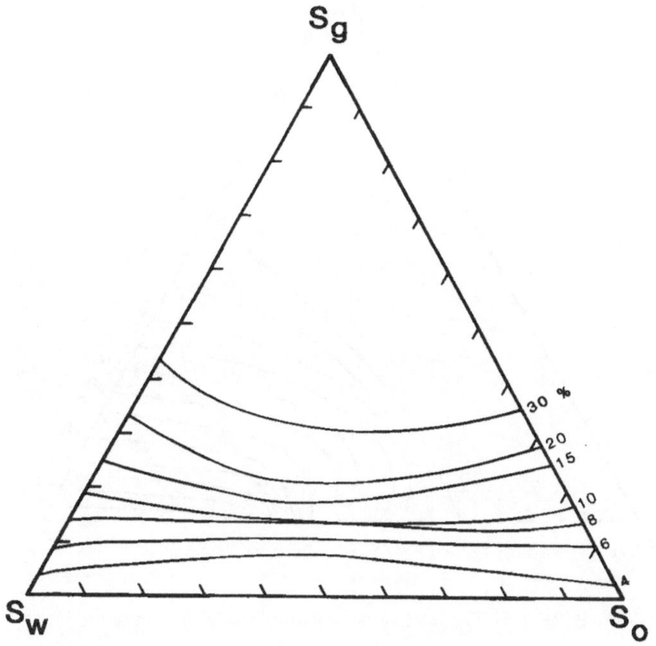

FIGURE 9. Gas relative permeability for Berea sandstone.[7]

The water isoperms are concave toward the water apex. Relative permeability to water was generally higher in the presence of oil than in the presence of gas, but k_{rw} was higher in the presence of gas than in its absence at a constant high S_w. Both k_{ro} and k_{rw} increased

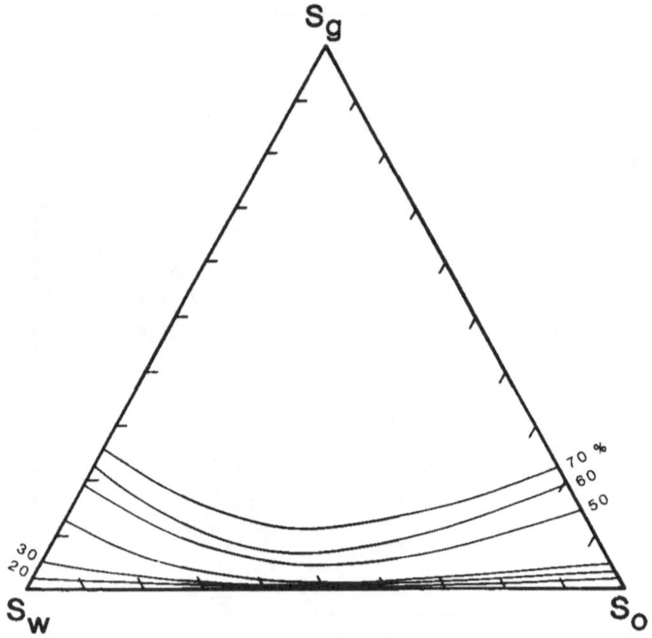

FIGURE 10. Gas relative permeability for Arbuckle limestone.[7]

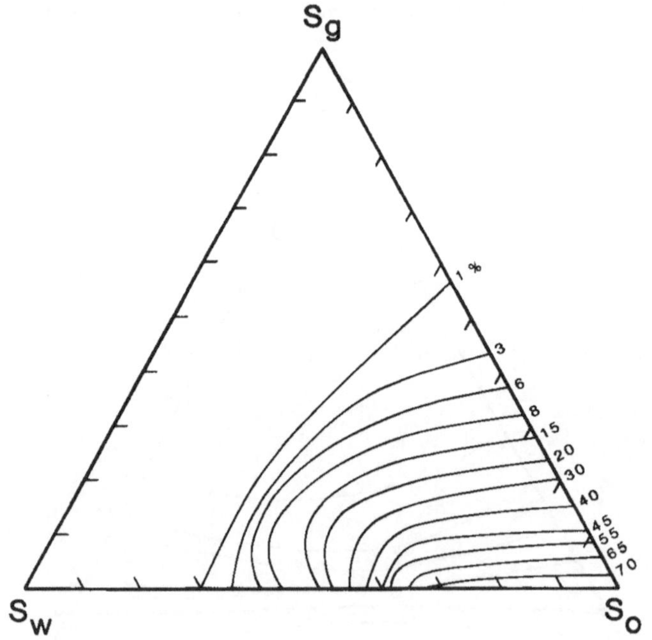

FIGURE 11. Oil relative permeability for Berea sandstone.[7]

at constant S_o and S_w, respectively, when S_g was increased from 0 to 8% possibly because gas was trapped in pores which would otherwise be occupied by immobile wetting phases. Also, k_{rw} increased in the presence of oil because there may be partial oil wetting, so that water was displaced into larger pores; this was not the case when gas was present.

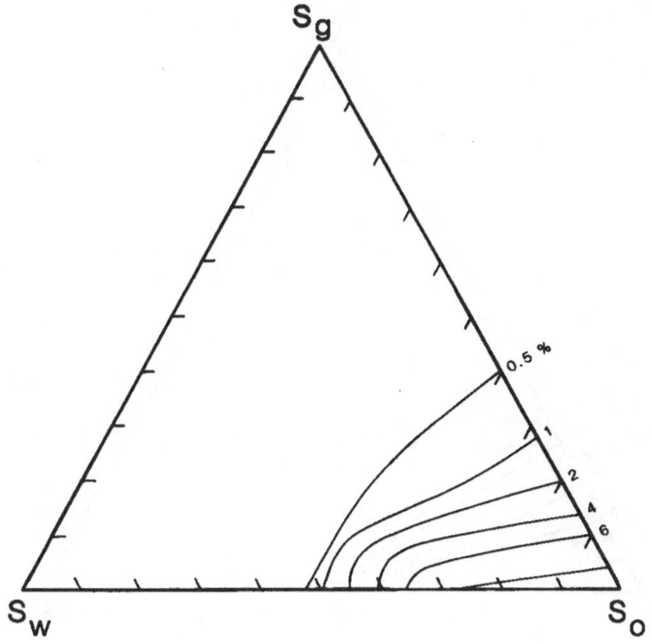

FIGURE 12. Oil relative permeability for Arbuckle limestone.[7]

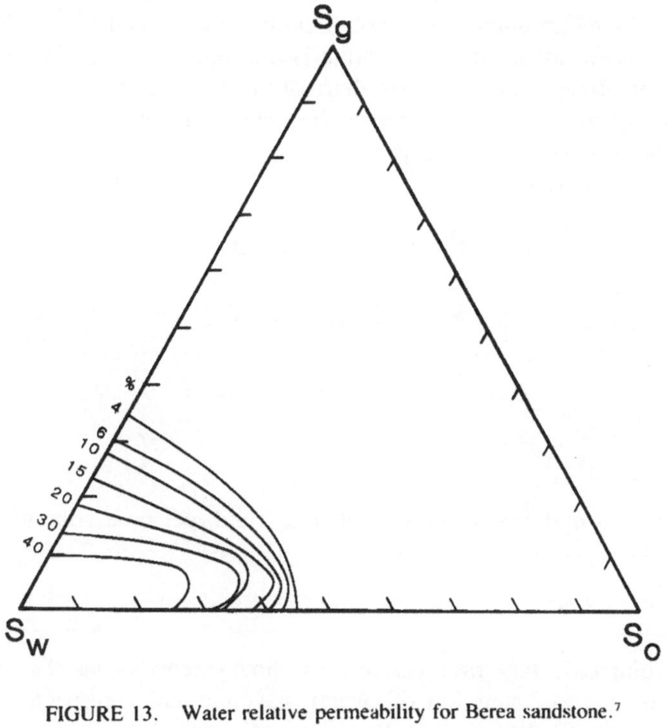

FIGURE 13. Water relative permeability for Berea sandstone.[7]

F. Sarem

Using an unsteady-state method, Sarem[8] obtained three-phase data for a Berea core. He did not consider end effects or saturation history, but his method did account for wettability. Sarem's method, which is an extension of Welge's two-phase technique, is relatively fast.

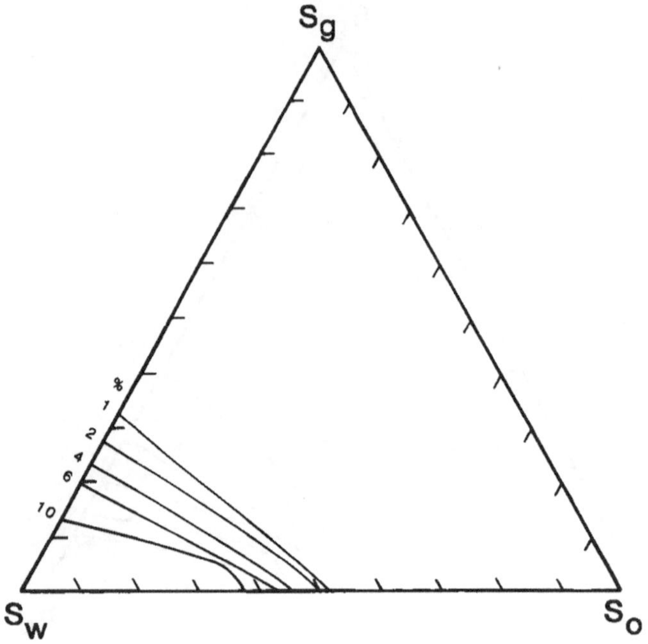

FIGURE 14. Water relative permeability for Arbuckle limestone.[7]

The core is saturated first with one liquid and then flooded with an immiscible unreactive liquid, at least until breakthrough. Then, both liquids are displaced by gas. In the derivation of his equations, Sarem assumed each relative permeability to be a function of the corresponding saturation alone. Isoperms were therefore parallel to the isosaturation lines. The relative permeability to gas was assumed to be dependent on total liquid saturation and independent of the relative wetting properties.

The saturation equations are

$$S_{o2} = S_{o,avg} + f_{o2}\,Q \tag{3}$$

$$S_{w2} = S_{w,avg} + f_{w2}\,Q \tag{4}$$

$$S_{g2} = 1 - S_{w2} - S_{o2} \tag{5}$$

where

Q = cumulative volume of injected fluid (per pore volume)

$$Q = \frac{q_T t}{LA\phi}$$

and q_T = total volumetric flow rate (cc/sec), t = time (seconds), and f = fractional flow.

Subscripts o, w, g, and 2 stand for oil, water, gas, and outlet, respectively. The relative permeabilities are computed from the following relationships:

$$k_{rw} = f_{w2}\frac{d\,(1/Q)}{d\left(\dfrac{\Delta PkA}{L\mu_w q_T Q}\right)} \tag{6}$$

$$k_{rg} = k_{ro} \frac{\mu_g f_g}{\mu_o f_o} \tag{7}$$

$$k_{ro} = f_{o2} \frac{d\,(1/Q)}{d\left(\frac{\Delta PkA}{L\mu_o\, q_T Q}\right)} \tag{8}$$

Sarem also concluded that initial saturation conditions affect k_{ro} and k_{rw}, but have little effect on k_{rg}. He found that k_{ro}/k_{rw} was influenced by initial saturations in three-phase studies in the same manner as in two-phase studies. Sarem's results differed from those of Donaldson and Dean even though both used the same type of sandstone.

G. Saraf and Fatt

A dynamic method using nuclear magnetic resonance (NMR) techniques was used by Saraf and Fatt[9] to determine liquid saturations in Boise sandstone. A volumetric method was used to obtain gas saturations. The experimental technique was designed to minimize end effects. To maintain a constant pressure differential, the gas flow rate was increased as the oil flow rate was decreased. Saraf and Fatt found no theoretical justification for Sarem's assumption that three-phase relative permeability to each phase was a function only of the saturation of that phase. In the water-wet Boise sandstone, however, they did find that k_{rw} was a function of S_w alone. Using water permeability as the base, they found that k_{rg} depended only on the total liquid saturation and was independent of the relative wetting properties. Oil isoperms determined by these investigators were convex toward the oil apex. Their results are shown in Figure 15. The explanation given by the authors for this unexpected shape of the isoperms seems less than convincing. They did state, however, that in studies where k_{rw} was a function of both S_w and S_o, the system was not 100% water-wet. In such a case, it seems likely that S_w did not remain constant when S_o or S_g was increased and that the assumption of constant S_w could be a source of experimental error.

H. Wyllie and Gardner

Three-phase relative permeability equations for preferentially water-wet systems where water and oil saturations were determined by the drainage cycle rather than by imbibition have been given by Wyllie and Gardner[10] and are presented in Chapter 2, Table 3. The following factors should be taken in consideration when using the equations presented into this table.

1. The k_{rw} values are normalized to absolute permeability.
2. The values of k_{ro} and k_{rg} calculated from these relationships are both normalized to the effective hydrocarbon permeability at irreducible water saturation. Inasmuch as they are normalized to the same base, k_{rg}/k_{ro} values may be calculated directly by using these equations. This is not true, of course, for water-hydrocarbon relative permeability ratios.
3. The gas and oil relative permeability equations do not include provision for residual oil saturation. When S_w equals S_{wirr}, k_{ro} is equal to $[S_o/(1 - S_{wirr})]^4$ for cemented sandstone, oolitic limestones, and vugular rocks. To handle residual oil saturation, this relationship should be altered to $[(S_o - S_{or})/(1 - S_{wirr})]^4$.

The correlations developed by Wyllie and Gardner can be used to construct a ternary diagram showing the relative permeabilities to oil, gas, and water. In general, the values of relative permeability (10, 20, 30%, etc.) are chosen first and then the values of saturation are obtained from the correlations. As can be seen from Chapter 2, Table 3, some of the

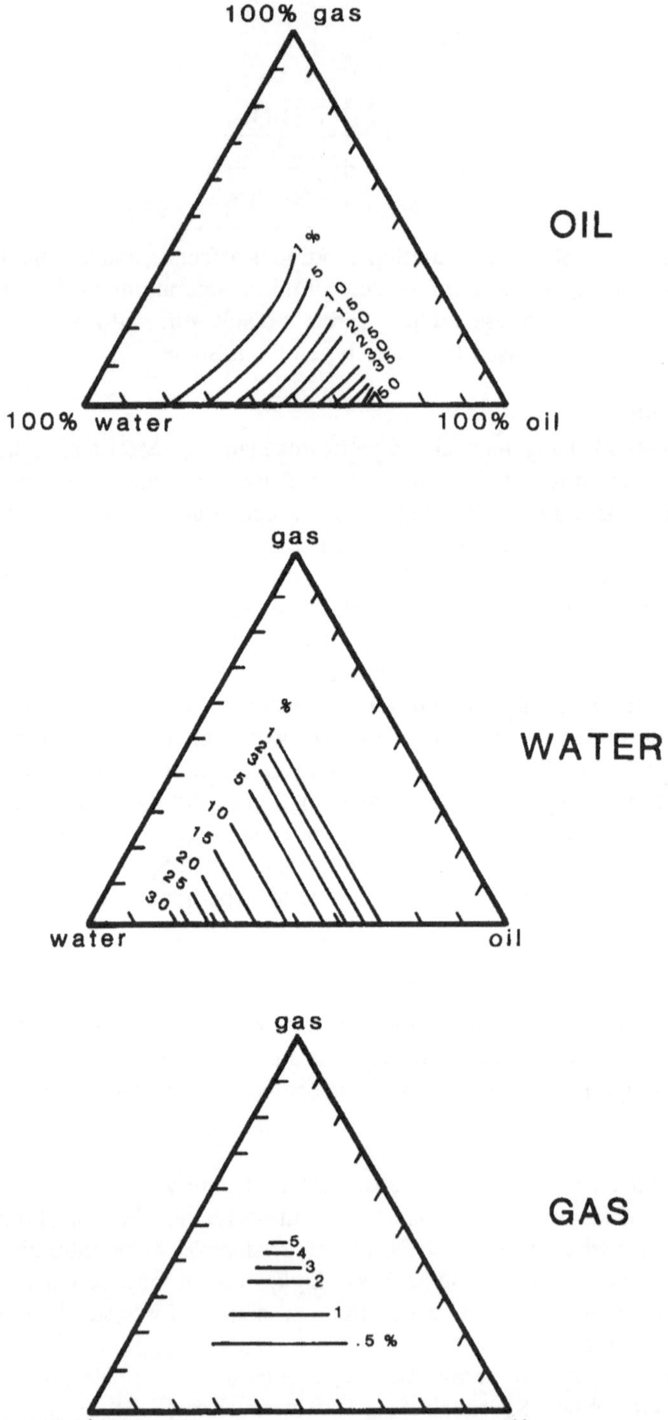

FIGURE 15. Three-phase relative permeability.[9]

equations are nonlinear. Hence, numerical methods (such as Newton-Raphson) are required to solve these equations. Manual interpolation is also possible for plotting relative permeability isoperms.

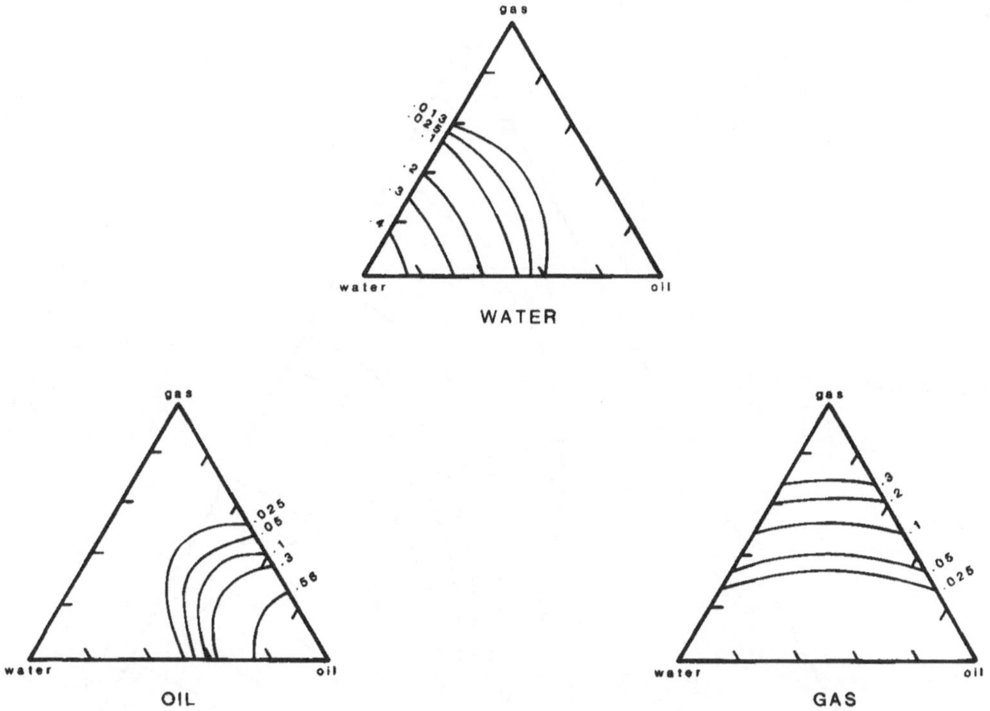

FIGURE 16. Three-phase relative permeability data of Caudle et al.[11]

Empirical relationships provide reasonable results in some cases and very disappointing ones in other situations; consequently, they must be used carefully. Note that most of the previous relationships were developed for media with intergranular porosity. This points out the huge problem of determining relative permeability curves for naturally fractured reservoirs. The difficulty arises primarily from the difficulty (or impossibility) of making this type of measurement on a fractured core sample.

For totally oil-wet three-phase systems in which oil is the wetting phase, water the nonwetting phase, and gas nonwetting with respect to both, the substitution of S_o for S_w in the Wyllie and Gardner equations can be made for estimation of the relative permeability to each phase.

III. IMBIBITION RELATIVE PERMEABILITY

A. Caudle, Slobod, and Brownscombe

Using a dynamic displacement method on a consolidated core sample, Caudle et al.[11] obtained isoperms for k_{ro}, k_{rw}, and k_{rg}, as shown in Figure 16. They used distillation to find the water and oil saturations at each data point, and used material balance for determination of gas saturation. Caudle et al. employed a pressure differential of 5 to 50 in. of water across the core and used water permeability as the base value. Relative permeability to water k_{rw} was found to be dependent on S_w, S_g, and S_o. These workers recognized the presence of some form of hysteresis in the three-phase studies, but they ignored the capillary end effect. They found all relative permeabilities to be approximately at minimum values when S_o was maintained at the value of S_{wc}.

B. Naar and Wygal

Naar and Wygal[12] developed a set of equations that was discussed in Chapter 2. Based

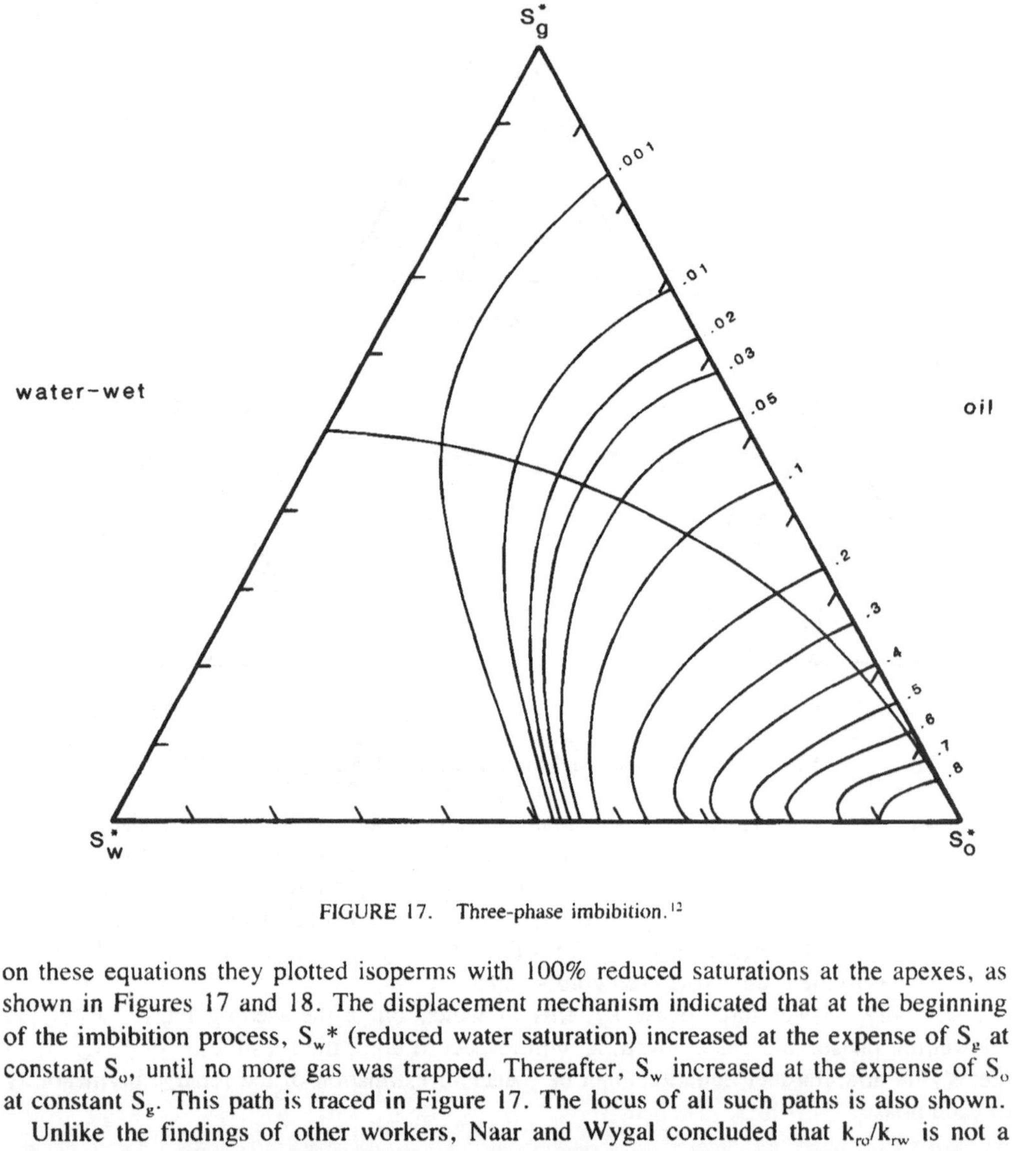

FIGURE 17. Three-phase imbibition.[12]

on these equations they plotted isoperms with 100% reduced saturations at the apexes, as shown in Figures 17 and 18. The displacement mechanism indicated that at the beginning of the imbibition process, S_w* (reduced water saturation) increased at the expense of S_g at constant S_o, until no more gas was trapped. Thereafter, S_w increased at the expense of S_o at constant S_g. This path is traced in Figure 17. The locus of all such paths is also shown.

Unlike the findings of other workers, Naar and Wygal concluded that k_{ro}/k_{rw} is not a function of S_{wi} for equal values of oil recovery in three-phase flow. On the other hand, the ratio was found to be a function of S_{gi} and wettability. This dependence is shown in Figure 19. The higher the initial gas saturation, the less the influence of wettability on k_{ro}/k_{rw}. Also, the water saturation at a given recovery was a function of initial water saturation and initial gas saturation. The ratio of S_{wi} for a water-wet system to S_{wi} for an oil-wet system increased with S_{gi}, and the rate of increase was an incresing function of S_{wi}. For a given S_w ratio and a given recovery, S_{wi} decreased with increasing S_{gi}. With higher S_{gi}, there is less pore space available and the oil is already pushed out into larger channels because of the higher S_{gi}; therefore, less water is required for the same recovery.

The imbibition water-oil relative permeability equations developed by Naar and Wygal, based on the assumption that $1/P_c^2$ equals CS*, are

$$k_{rw,imb} = \frac{S_{wi^*,imb} \displaystyle\int_0^{S_{w^*,imb}} \frac{S_{w^*,imb} - S^+ \, dS^+}{P_c^2}}{\displaystyle\int_0^1 \frac{1 - S^+ \, dS^+}{P_c^2}} \tag{9}$$

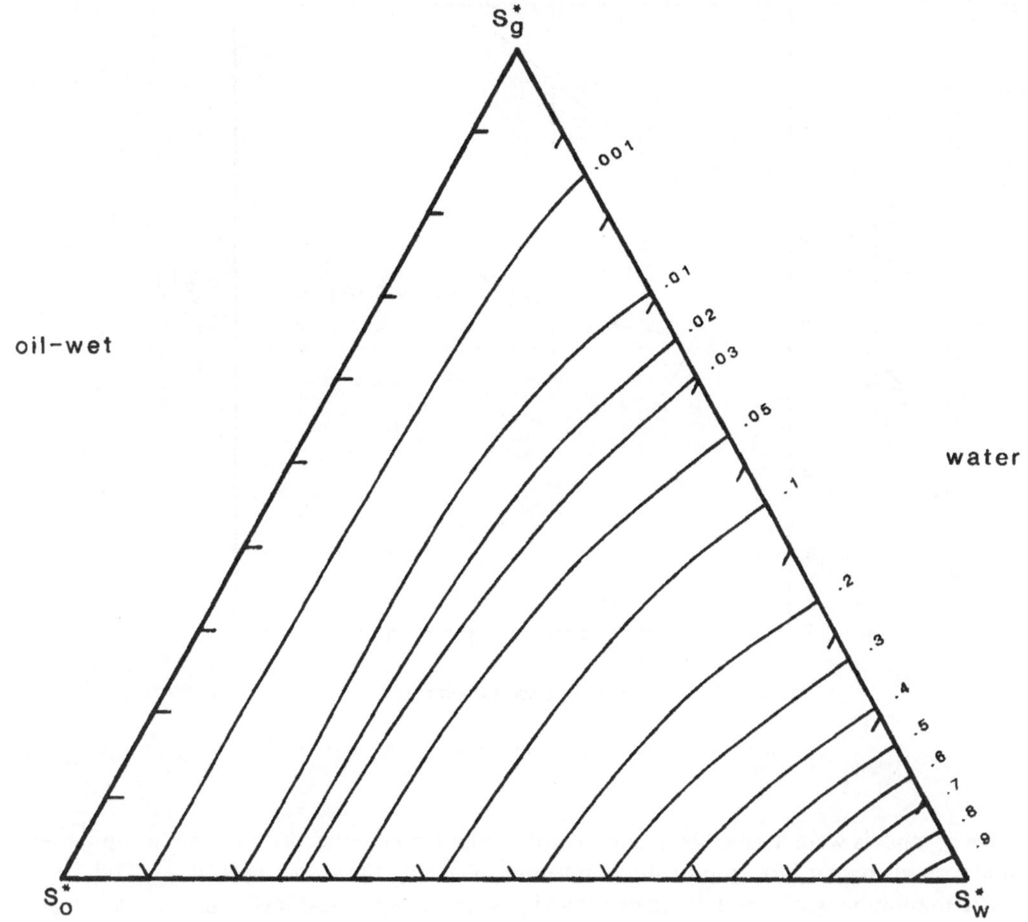

FIGURE 18. Three-phase drainage.[12]

and

$$k_{ro,imb} = S_{of}^{*3}\,(S_{of}^{*} + 3\,S_{fw}^{*})$$
$$= \frac{(S_o - S_{ob})^3\,(S_o + 2S_{ob} + 3S_w - 3S_{wi})}{(1 - S_{wi})^4} \qquad (10)$$

Naar and Wygal suggested the following approximation for imbibition gas relative permeability:

$$k_{rg} = \frac{0.5 - S^*_{wi,imb}}{0.5}\,(1 - S^{*2}_{w,imb}) \qquad (11)$$

where

$$S^*_{w,imb} = S^*_{w,drain} - 1/2\ S_w^{*2},\ \text{drain};$$

$$S^+ = \frac{S - S_{wi}}{1 - S_{wi}}$$

$$S_{of}^* = \frac{S_o - S_{ob}}{1 - S_{wi}}$$

$$S_{fw}^* = \frac{S_w - S_{wi} + S_{ob}}{1 - S_{wi}}$$

and S_{ob} is the trapped oil saturation.

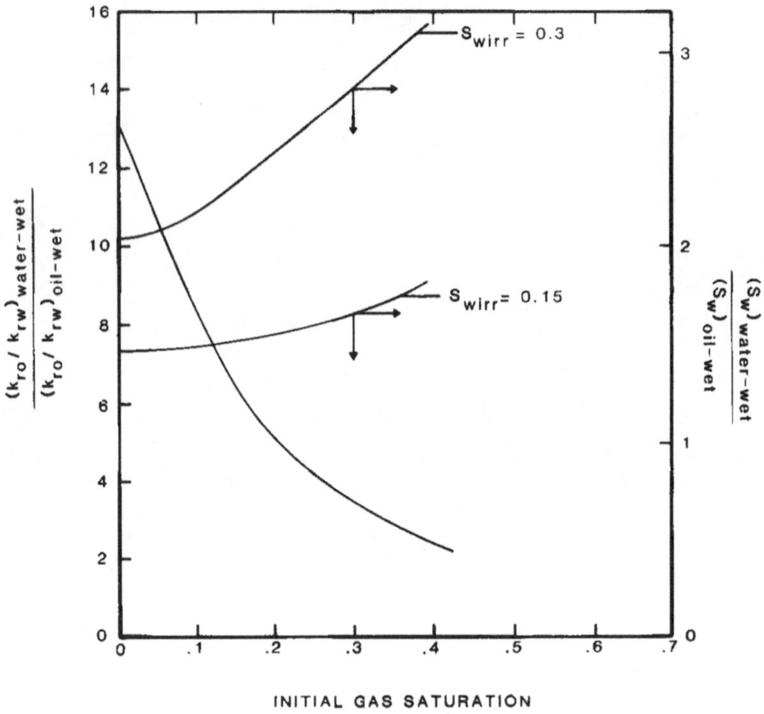

FIGURE 19. Influence of wettability at 40% recovery.[12]

These models were derived by assuming the random interconnection of straight capillaries, with a provision for blocking of the nonwetting phase by the invading wetting fluid.

The imbibition water, and drainage oil and gas, relative permeability equations developed by Naar and Wygal were also presented in the following form:

$$k_{rw,imb} = (S_w^*)^4 \quad (12)$$

$$k_{ro,imb} = (1 - 2S_w^*)^{3/2} \{2 - (1 - 2S_w^*)^{1/2}\} \quad (13)$$

and

$$k_{rg} = S_{gf}^{*3} (3 - 2S_{gf}^*) \quad (14)$$

where

$$S_{gf}^* = \frac{S_g - S_{gt}}{1 - S_{wi}}$$

In these equations the subscript "t" stands for "trapped" and "f" for "free".

C. Land

In Land's[13] work, equations for imbibition two- and three-phase relative permeabilities were obtained from rock properties. Land considered residual gas saturation after imbibition to be directly related to the initial gas saturation. The gas and water imbibition relative permeabilities were reported to be the same in three-phase systems as in two-phase systems,

OIL

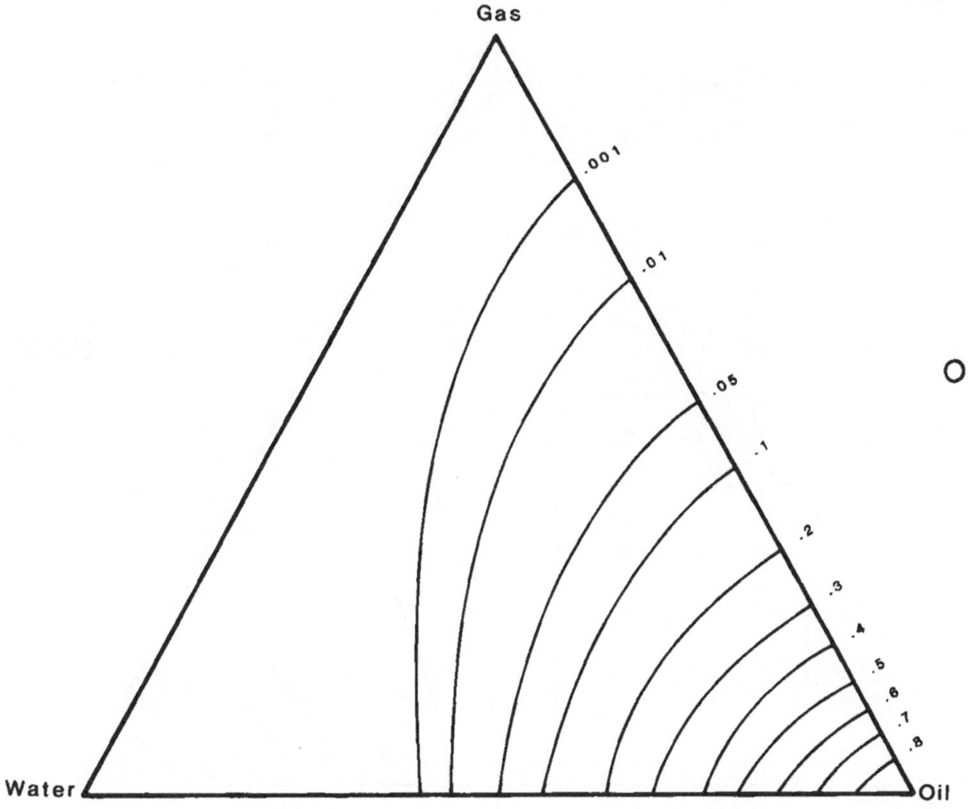

FIGURE 20. Imbibition k_{ro} for a mobile gas saturation.[13]

since the totally nonwetting and wetting phases occupied the same pores regardless of the nature of the other phases present.

His plots for k_{ro} in the II and ID cases are shown in Figures 20 and 21. The ID plots are similar to the plots obtained by Naar and Wygal,[12] their system being an II case.

Land's final equations are

$$k_{rg} = \frac{S_{gf}^{*2} \int_{1-S_{gf}}^{1} \frac{dS^*}{P_c^2}}{\int_0^1 \frac{dS^*}{P_c^2}} \tag{15}$$

$$k_{rw} = \frac{S_w^{*2} \int_0^{S_w^*} \frac{dS_w^*}{P_c^2}}{\int_0^1 \frac{dS^*}{P_c^2}} \tag{16}$$

$$k_{ro} = \frac{S_{of}^* \int_{S_w^* + S_{ot}^*}^{S_w^* + S_o^*} \frac{dS_L^*}{P_c^2}}{\int_0^1 \frac{dS^*}{P_c^2}} \tag{17}$$

For S_w increasing and S_g constant:

$$k_{ro} = S_{of}^{*3}\,[2(S_w^* + S_{ot}^*) + S_{of}^*] \tag{18}$$

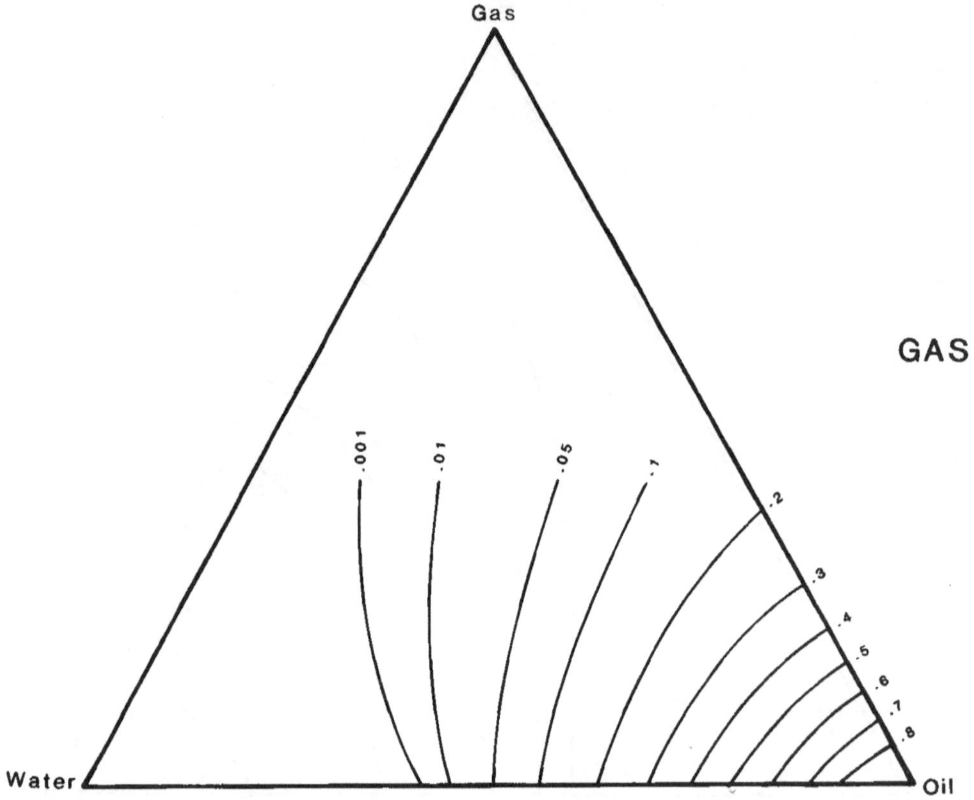

FIGURE 21. Imbibition k_{ro} for a trapped gas saturation.[13]

This equation is similar to the one obtained by Corey et al.[2] for the drainage condition. When all the gas is trapped:

$$k_{ro} = S_{of}^{*3}(2S_w^* + S_{of}^*) - S_{of}^*[S_{gr}^{*2} + 2/C(S_{gt}^* + 1/C\{\ln S_{gt}^*/S_{gt}^*\})] \quad (19)$$

where

$$S^* = (1 - S_{gf}^*)$$

$$S_o^* = \frac{S_o - S_{om}}{1 - S_{wi} - S_{om}}$$

$$S_{gt}^* = \frac{S_{gt}}{1 - S_{wi}}$$

$$S_{ob}^* = \frac{S_{ob}}{1 - S_{wi}}$$

$$S_{of}^* = \frac{S_o - S_{ob}}{1 - S_{wi}}$$

$$C = \frac{1}{(S_{gt}^*)_{max}} - 1$$

$$S_w^* = \frac{S_w - S_{wc}}{1 - S_{wc}}$$

S_{om} = minimum residual oil saturation

S_{gt} = trapped gas saturation

S_{ob} = trapped oil saturation

Land's correlations did not consider hysteresis since his derivation was based on the work of Corey et al., which did not include hysteresis effects.

D. Schneider and Owens

Schneider and Owens[14] performed steady-state and unsteady-state tests on a variety of carbonate and sandstone samples, and found the relative permeability to oil during an imbibition process in a water-wet system to be insensitive to the flowing gas phase when gas saturation was increasing. Oil relative permeability was found to be primarily dependent on oil saturation. It was reported that residual oil significantly reduced the gas relative permeability in a water-wet system. The gas relative permeability in an oil-wet system was found to be insensitive to the presence of a residual oil saturation. The nonwetting relative permeability-saturation relationship in three-phase flow was reported to depend on the saturation history of both nonwetting phases and on the ratio of the saturations of the two wetting phases. In some cases the nonwetting relative permeability was found to be lower then the two-phase value due either to trapping of a nonwetting phase or to flow interference between the nonwetting phases when both were mobile. For some tests the nonwetting relative permeability value for three-phase flow was found to be higher than the two-phase value. The authors discussed the reasons why their results did not fully agree with those of Corey et al.

E. Spronsen

The centrifuge method, already proven for two-phase flow, was extended by Spronsen[15] to drainage three-phase flow in a water-wet system. Oil isoperms determined by Spronsen are concave toward the 100% oil apex. He discussed the adverse influence of immiscible CO_2 injection on the shape of three-phase oil isoperms. The results of his investigation are shown in Figure 22.

IV. PROBABILITY MODELS

Since the experimental problems associated with three-phase flow are difficult to surmount, a mathematical model appears to be an alternate approach. The correlations discussed earlier required some type of experimental three-phase flow data. On the other hand, probability models as formulated by Stone[16,17] and modified by Dietrich and Bondor[18] and later by Nolen as cited by Molina,[19] assume that two-phase flow behavior can be used as a limiting condition for three-phase flow. Water-oil-gas flow can be bounded by water-oil flow at one extreme and oil-gas flow at the other. While some of these models can consider hysteresis, the earlier correlations, such as Land's,[13] cannot do so.

Most probability models assume that gas relative permeability is dependent only on the gas saturation:

$$k_{rg} = k_{rg}(S_g) \tag{20}$$

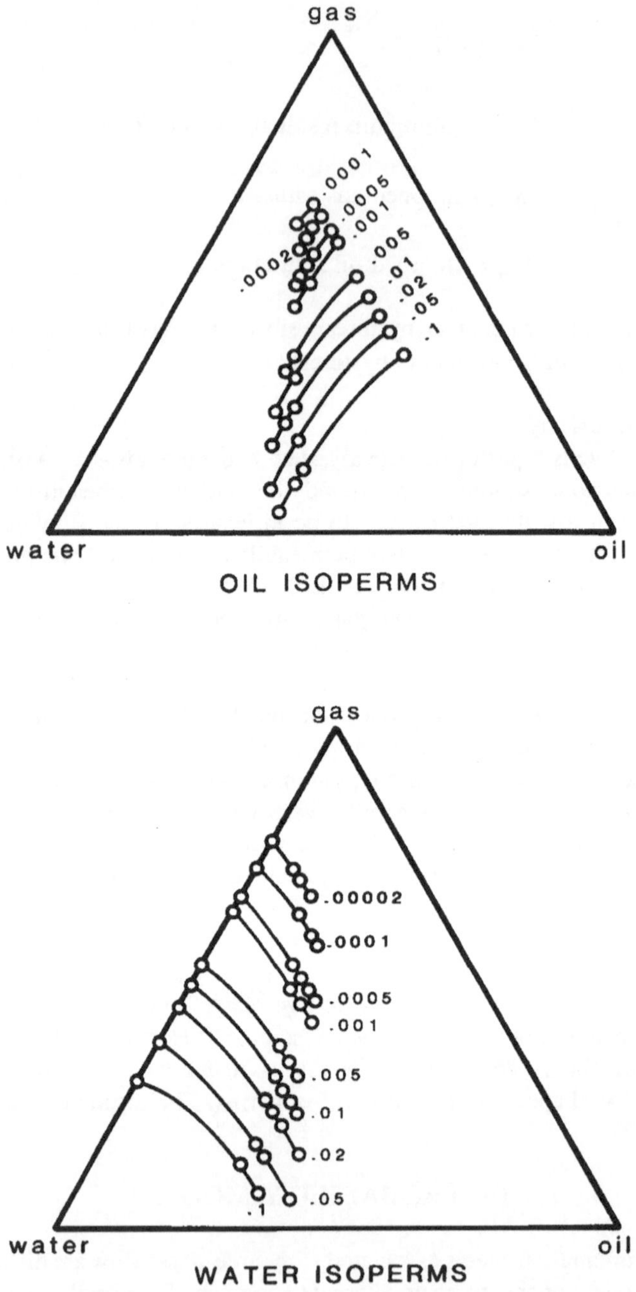

FIGURE 22. Data of Spronsen for Berea sandstone.[15]

Similarly, it is assumed that the relative permeability to water is dependent only on the water saturation:

$$k_{rw} = k_{rw}(S_w) \tag{21}$$

Oil relative permeability, however, varies in a more complex manner. These assumptions have been confirmed in laboratory investigations for a water-wet system.

In a water-wet system, gas behaves as a completely nonwetting phase, but oil has an intermediate ability to wet the rock. The relative permeability to oil in a water-oil-gas system will therefore be bounded by relative permeability to oil in a water-oil system at low gas saturations and by relative permeability to oil in a gas-oil system at low water saturations. Stone attempted to combine these two terminal relative permeabilities to obtain a three-phase result by using the channel flow theory in porous media and simple probability models. Water and gas three-phase relative permeabilities, according to Stone, are the same as their corresponding two-phase relative permeabilities. In his first model, Stone developed the expression:

$$k_{ro} = S_o^* \beta_w \beta_g \tag{22}$$

where

$$S_o^* = \frac{S_o - S_{or}}{1 - S_{wi} - S_{or}}$$

$$\beta_w = \frac{k_{row}}{1 - S_{wr}} \quad (2-\text{phase})$$

$$S_{wr}^* = \frac{S_w - S_{wi}}{1 - S_{wi} - S_{or}}$$

$$\beta_g = \frac{k_{rog}}{1 - S_g^*} \quad (2-\text{phase})$$

and

$$S_g^* = \frac{S_g - S_{gc}}{1 - S_{wi} - S_{or} - S_{gc}}$$

Fayers and Matthews[26] suggested that

$$S_{or} = \alpha\, S_{orw} + (1 - \alpha)\, S_{org}$$

where

$$\alpha = 1 - \frac{S_g}{1 - S_{wc} - S_{org}}$$

Stone's earlier model did not agree well with data involving the dependence of waterflood residual oil saturation on trapped gas saturations. Stone's second model gave three-phase oil relative permeability as

$$k_{ro} = (k_{row} + k_{rw})(k_{rog} + k_{rg}) - (k_{rw} + k_{rg}) \tag{23}$$

where k_{row} and k_{rw} represent oil and water relative permeabilities from two-phase, oil-water relative permeability data; k_{rog} and k_{rg} represent oil and gas relative permeabilities from two-phase, oil-gas relative permeability data. Equation 23 may yield unrealistic results at low k_{ro} values.

Although it seems reasonable that one should be able to combine the two two-phase relative permeabilities to arrive at three-phase data at least for water-wet systems, the manner in which they have been combined in these models may not account for the total physics of the process. These probability models strongly depend on the assumption that there is at most one mobile fluid in any channel. That is, Stone's assumption implies that water-oil capillary pressure and water relative permeability are functions of water saturation alone in the three-phase system, regardless of the relative saturations of oil and gas. Moreover, they are the same function in the three-phase system as in the two-phase gas-oil system. Stone's second model generally predicts the correct oil relative permeability in the three-phase system if the relative permeability at the end points is equal to one. Stone points out that when his second model yields a negative k_{ro}, this implies a complete blockage of oil and as a result k_{ro} equals zero. The Stone models account for hysteresis when water and gas saturations are changing in the same direction.

Dietrich and Bondor[18] applied Stone's models to published three-phase data and found them to be only partially successful. They found that it was necessary to modify Stone's second model for the case where gas/oil relative permeability is measured in the presence of connate water. They pointed out that, at irreducible water saturation and zero gas saturation, Equation 23 reduced to

$$k_{ro} = (k_{row})(k_{rog})$$

This expression can be valid only if both k_{row} and k_{rog} equal unity. Since k_{ro} at S_{wc} is frequently less than one, Stone's second model has some limitations.

Dietrich and Bondor adjusted Stone's model by normalizing it with k_{rocw} to obtain:

$$k_{ro} = \frac{1}{k_{rocw}} [(k_{row} + k_{rw})(k_{rog} + k_{rg})] - (k_{rw} + k_{rg}) \tag{24}$$

where k_{rocw} is the oil relative permeability at connate water saturation. At irreducible water saturation and zero gas saturation this equation reduces to:

$$k_{ro} = \frac{(k_{row})(k_{rog})}{k_{rocw}}$$

This model tends to predict incorrect oil relative permeability values (magnitude larger than unity) for values of $k_{rocw} \leqq 0.3$.

Nolen, as referenced by Molina[19] has taken into account this problem and suggested the following model which remains bounded as k_{rocw} approaches zero:

$$k_{ro} = k_{rocw} \frac{k_{row}}{k_{rocw}} + k_{rw} \frac{k_{rog}}{k_{rocw}} + k_{rg} - (k_{rw} + k_{rg}) \tag{25}$$

V. EXPERIMENTAL CONFIRMATION

Three-phase relative permeability studies are still in an early stage of development. Little has been done on the experimental confirmation of imbibition correlations and most of the correlations available are for imbibition.

Early work was done primarily on unconsolidated sands and the effects of wettability and hysteresis were not recognized until recently. Donaldson and Kayser[20] have categorized the reasons for divergence of experimental three-phase relative permeability data as follows:

1. Errors introduced in saturation measurements in various experimental methods.
2. Errors introduced by neglect of capillary end effects and saturation hysteresis phenomena.
3. Variations caused by use of different oils, brines, and cores which could exhibit different wettability characteristics.
4. Assumptions made to facilitate experimental procedures or calculations.
5. Inadequacy of mathematical formulations to represent three-phase flow conditions.

The empirical methods, though seemingly simpler, suffer from simplifying assumptions that have limited the range of saturation histories that can be simulated.

Table 1 is a chronological listing of the experimentally determined three-phase relative permeabilities that have been reported.[21] In all of the studies included in the tabulation the authors used refined oils in order to minimize oil-wetting; they assumed a totally water-wet system. In cases where a single core was used, the influence of the saturation history of the rock sample was frequently ignored. The gases used in the studies listed in Table 1 were air, carbon dioxide, and nitrogen.

VI. LABORATORY APPARATUS

Three-phase relative permeability studies have been conducted using refined nonpolar oil, diesel oil, Soltrol, kerosene, hydrocarbon fractions, brine, nitrogen, air, and carbon dioxide. Berea, Boise, Torpedo, Tensleep, and Weeks Island sandstones, as well as Arbuckle limestone and unconsolidated sand samples have been used for the flow media. Berea sandstone is often preferred because of its uniformity and general acceptability as an industry standard.

Personal communications with researchers in this field have indicated that the most practical means of saturation measurement is gravimetric. Other modern methods, such as gamma ray absorption, X-ray absorption, NMR, etc., are unnecessarily expensive and elaborate. The gravimetric saturation measurements are sufficiently accurate and relatively inexpensive. Problems may be encountered with gravimetric saturation measurements, however, especially when gas is used in the presence of volatile oil. Therefore, core holders which permit rapid removal of cores (without the removal of rubber sleeves) should be used when relative permeability is determined by steady-state methods. Wettability of the core should be monitored either by the centrifugal technique[23] or an alternative method.

Brine saturation may be determined with satisfactory accuracy by electrical resistivity measurement when nonpolar oil is employed. Oil saturation may be obtained gravimetrically and the gas volume may be computed as the difference between total pore volume and total liquid volume. The oil and water flow rates may be obtained by a simple burette arrangement or by flowmeters. The gas flow rate may be obtained by use of a gas flowmeter.

The effect of wettability on the relative permeabilities is an important factor that should be studied. The change of wettability in a core from oil-wet to a water-wet has been known to influence relative permeabilities, but no definite conclusions are found in the literature concerning the influence of wettability on three-phase relative permeabilities.

Boundary effects should be eliminated by using core plugs at either end of the test core and performing the experiment at reasonably high flow rates. A semipermeable membrane may precede the core plug at the inlet end for proper distribution of the phases. A modified Penn State method of relative permeability measurement may be used, since most investigators believe that the Penn State method gives better results than any of the other steady-state methods.

In addition to saturation measurements for each phase, one needs to measure the flow rate of each fluid and the pressure drop while making the steady-state relative permeability measurements. A gas dome may inject fluids into the core and a back-pressure regulator may be used to maintain a constant pressure at the outlet end. Also, the gaseous phase should be bubbled through the oil supply tanks. This procedure ensures that the oil is saturated with the gas before it enters the core. As a result, there should be no mass transfer between the gaseous phase and the oil inside the core. The gas flow rate may be regulated with a needle valve, with a large pressure differential across the valve. The rate of gas flow may be measured either by collecting gas by displacement of water for a known length of time, or with a soap bubble meter or a wet-test meter. For pressure differential measurements, it is desirable that the displacement of fluid into the measuring device be as small as possible so as to minimize error. Hence, use of a regular manometer is not possible, but differential pressure transducers may be used. The connections for measuring pressure differential can

Table 1
CHRONOLOGICAL LISTING OF THREE-PHASE RELATIVE PERMEABILITY INVESTIGATIONS[21]

Authors	Date	Materials	Experimental methods	Saturation measurement techniques	Treatment of hysteresis effects	Treatment of end-effect	Number of measurements	Relative permeability is a function of		
								k_{rw}	k_{ro}	k_{rg}
Leverett and Lewis	1941	Unconsolidated sand (φ = 42%; k = 5 – 16 D) Gas = nitrogen Oil = kerosene Water = 0.24 *N* NaCl	Single core dynamic method	Water saturation from resistivity measurement	Ignored	Neglected	64	S_w	S_o, S_w, S_g	S_g, S_o, S_w
				Gas from pressure-volume measurement Oil from material balance		Errors incurred from low flow rates				
Caudle, Slobod and Brownscombe	1951	Consolidated sandstone (φ = 23%; k = 0.025D) Gas = air	Penn State method	Saturations determined for each point by removing the core and subjecting it to vacuum distillation	Ignored	Neglected	Unknown	S_w, S_o, S_g	S_o, S_w, S_g	S_g, S_o, S_w
Corey, Rathjens, Henderson and Wyllie	1956	Berea sandstone Water = $CaCl_2$ brine	Hasseler's capillary pressure method	Semipermeable diaphram assemblies were used at each end of the core to keep the water phase in the core. Gravimetric methods were used to determine fluid saturations	Hysteresis eliminated by using different core specimens for each saturation point	Minimized with the diaphrams	62	S_w	S_o, S_w, S_g	S_g

Reid	1956	Unconsolidated sand (ϕ = 35%; k = 31 D) Gas = air, Oil = diesel Water = 0.1 *N* NaCl	Single core dynamic method	Water by resistivity measurement	Ignored	Eliminated by making saturation and pressure measurements away from the ends of the core	98	S_w,S_o,S_g	S_o,S_w,S_g	S_g,S_o,S_w
Snell	1962	Unconsolidated sand (ϕ = 35%; k = 7.1 D) Gas = air, Oil = diesel, Water = 0.1 *N* NaCl	Penn State method	Water from the resonance behavior of an RCL circuit at radio frequency	Evaluated results independently from each other on the basis of saturation history	Eliminated by making measurments away from the ends of the core	250	S_w,S_o,S_g	S_o,S_w,S_g	S_g,S_o,S_w
Sarem	1965	Berea sandstone (ϕ = 20%; k = 0.275 D) Gas = air, Oil = Soltral 130, Water = 0.2 *N* NaCl	Extension of Welge's unsteady-state method	Volumetric measurements of terminal flow rates followed by calculation of terminal saturations using extended Welge method	Ignored	Neglected	70	S_w	S_o	S_g
Donaldson and Dean	1966	Berea sandstone (ϕ = 19%); Arbuckle limestone (ϕ = 11%; k = 0425D) Gas = air, Oil = Soltrol, Water = 0.1 *N* NaCl	Extension of Welge's 2-phase unsteady-state method to 3-phase	Volumetric measurements of terminal flow rates followed by calculation of terminal saturations using extended Welge method	Ignored	Minimized by using high flow rates, rates high as 0.5 PV/min	120	S_w,S_o,S_g	S_o,S_w,S_g	S_g,S_o,S_w
Saraf	1967	Boise sandstone (ϕ = 25%; k = 1.45 D) Gas = nitrogen, Oil = kerosene, Water = deuterium oxide	Single core dynamic method	Nuclear Magnetic resonance technique	Eliminated by using different core samples; no saturation was established twice in the same core	Eliminated by measuring saturation and pressure away from the ends of the core	81	S_w	S_o,S_w,S_g	S_g

Table 1 (continued)
CHRONOLOGICAL LISTING OF THREE-PHASE RELATIVE PERMEABILITY INVESTIGATIONS[21]

Authors	Date	Materials	Experimental methods	Saturation measurement techniques	Treatment of hysteresis effects	Treatment of end-effect	Number of measurements	Relative permeability is a function of		
								k_{rw}	k_{ro}	k_{rg}
Schneider and Owens	1970	Torpedo sandstone (ϕ = 24%: k = 370 mD)	Dynamic	Gas by X-ray absorption, Water by electrical resistivity	Evaluated results for oil-wet and water-wet systems, independently	Not discussed	Unknown	S_w	S_o, S_w, S_g	S_g, S_o, S_w
		Tensleep sandstone (oil-wet) (ϕ = 11.5%; k = 104 mD) Gas = nitrogen Oil = H-C fractions Water = brine								
Spronsen	1982	Berea sandstone (ϕ = 21.3% k = 480 — 600 μm^2) Weeks Island (ϕ = 26%, k = 3.57 - 4.3 μm^2) Gas = CO^2 Oil = unknown Water = water and glycol	Steady-state centrifuge	Gravitational	Steady-state oil-water imbibition always established	Eliminated by using high flow rates	Unknown	S_o, S_w, S_g	S_o, S_w, S_g	

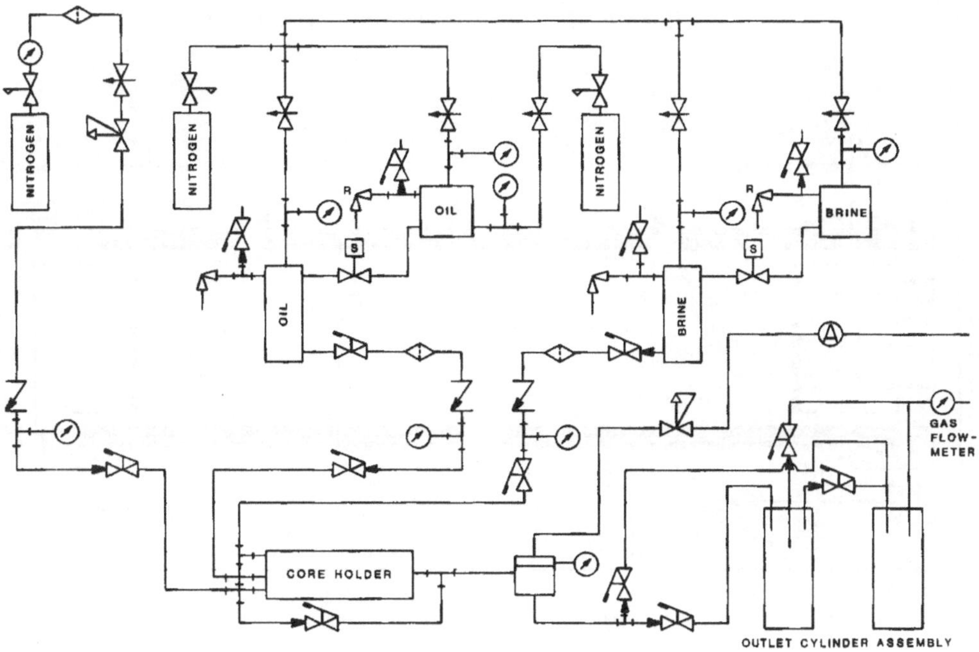

FIGURE 23. Schematic diagram of three-phase relative permeability apparatus.

be made through semipermeable membrane ports. The capillary tubes connecting the transducer may be inserted and cemented in place about 1 in. from each end of the test apparatus.

Unsteady-state methods of three-phase relative permeability measurements have the advantage of being rapid. Oil and water may be displaced by gas to duplicate gas drive processes used in enhanced recovery methods. However, the calculation of isoperms from laboratory data requires analytical solutions of the partial differential equations describing the three-phase fluid flow. Some early studies have made erroneous simplifying assumptions in describing the dynamic condition of the unsteady-state process. Reliable values of relative permeability as a function of saturations may be obtained by mathematical simulation of laboratory data using finite difference calculations.[20] Capillary pressure data should be obtained for gas-oil, water-oil, and water-gas systems to provide basic data necessary for three-phase relative permeability calculations. Solubility of the gas in the liquids employed in the study should be determined before these calculations are performed.

A schematic diagram of the apparatus used for three-phase relative permeability measurement is shown in Figure 23. The core holder, which has ports for differential pressure measurements, allows rapid retrieval of the core. Temperature is controlled with a Proportional Controller connected to a heating tape wrapped around the core holder. In order to eliminate pulsation of flow normally associated with pumps, fluids are injected by applying gas pressure on top of the fluid in a tank equipped with appropriate relief valves. Solenoid valves and level controllers maintain a constant head of fluid in the supply tanks. Filters are provided in the supply lines of each phase being injected into the core holder. Check valves prevent backflow of each of the three phases. A cross section of the core holder is shown in Figure 24.

Auxiliary equipment includes an accurate balance, electrical resistivity measurement system, level controller, chart recorder, differential pressure transducer, cylinders, compressed air and regulators, and a humidity oven.

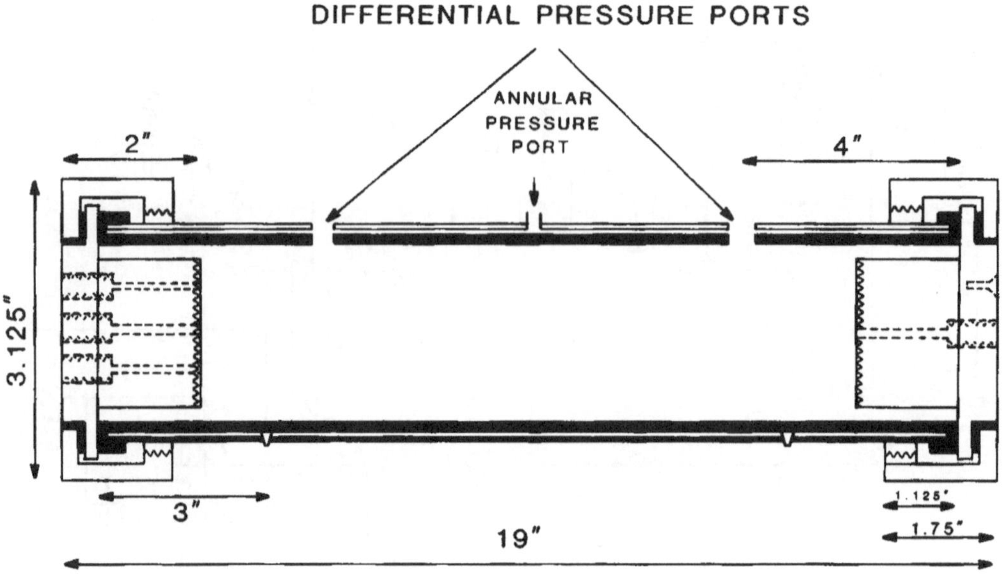

FIGURE 24. Diagram of a core holder.

VII. PRACTICAL CONSIDERATIONS FOR LABORATORY TESTS

The literature cited contains a large amount of information on factors affecting the laboratory investigation of relative permeability. The following listing, however, cites some practical considerations that have not been widely discussed in the literature:

1. If a pump is used to inject fluids into the core, the packing material should preferably be Teflon®. Most other packing materials contain silicon and carbon which may dissolve in injected fluids and affect the wettability of the core.
2. When brine is used as one of the fluids, all metal parts of the system should be of stainless steel. One-eighth-in. tubing offers excellent handling characteristics. Tygon tubing is recommended if the pressure is not too high.
3. Most electronic differential pressure transducers have good linearity and hysteresis characteristics; however, if possible, the transducer should be recalibrated at least once per month.
4. While changing pressures on the liquid storage tanks, it is important not to exceed the backpressure rating of the solenoid valves.
5. Every effort should be made to ensure 100% saturation of the wetting phase before starting injection of the nonwetting phase.
6. In a steady-state experiment, input flow rate should equal the output flow rate for each phase. In many cases, this condition is tedious to achieve.
7. Some extraneous material may be noticed in the output lines. It must be determined whether the particles are fines from the test sample or bacterial matter. A bactericide may be used with caution not to alter either the wettability or the resistivity of the core.
8. Often the resistivity meter utilizes chamois leather contacts at either end of the core holder. The contacts should be kept immersed in brine to prevent changes in the readings.
9. It has been noticed that the position of the outlet tubes going into the measuring cylinders affects the pressure differential readings. It is recommended that the tubing outlet be kept at the same level as the core holder to eliminate gravitational effects.

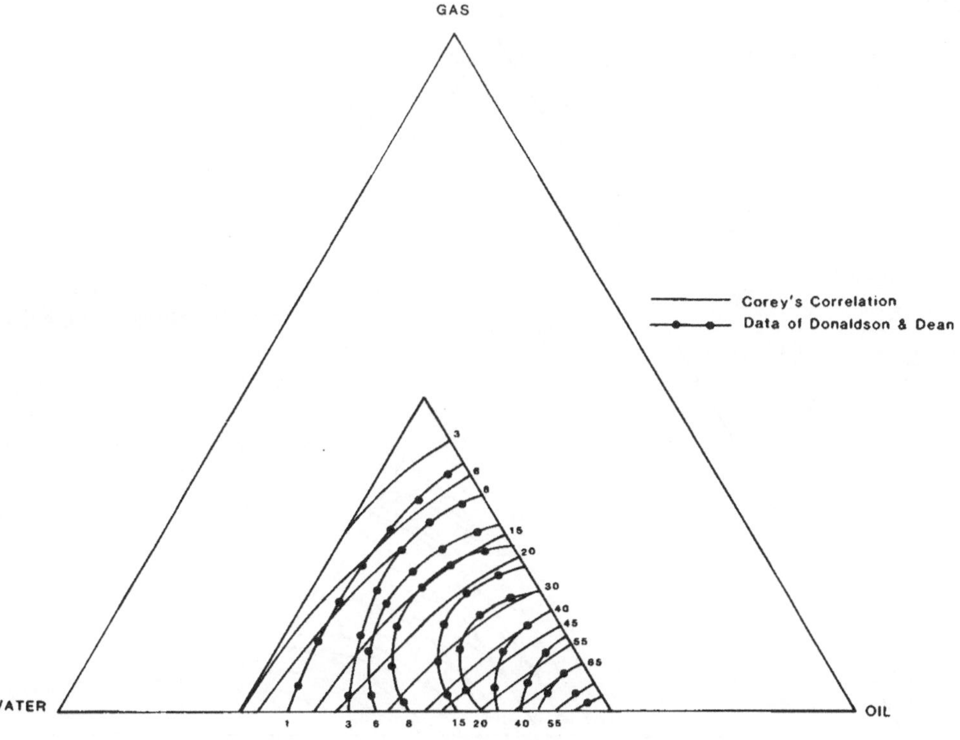

FIGURE 25. Comparison of three-phase oil relative permeability determinations.

10. Gas in the transducer lines seriously affects pressure differential readings. The transducer should be bled of gas at frequent intervals.
11. Every effort should be made to eliminate end effects as described by Batycky et al.[22]
12. If possible, the wetting characteristics of the core should be frequently monitored during the relative permeability experiments. The centrifuge method[23] may be employed for monitoring wettability.

VIII. COMPARISON OF MODELS

The following section presents a comparison of some of the models discussed earlier.

The equation of Corey et al.[2] for three-phase k_{ro} values is valid for a system in which oil is displaced by a gas. Donaldson and Dean[7] obtained three-phase k_{ro} values following the same displacement mechanism. Thus, we have an opportunity to observe how well the equation of Corey et al.[2] fits data provided by other workers. Three-phase oil relative permeability values calculated by the equation of Corey et al.[2] were compared with Donaldson and Dean's data. The isoperms obtained are shown in Figure 25 along with Donaldson and Dean's data as a basis for comparison. The Corey et al.[2] equation gives higher k_{ro} values than those obtained by Donaldson and Dean. Isoperms by Corey et al.[2] are less concave towards 100% oil saturation. Both methods are in agreement in predicting that the isoperms become concave toward 100% S_o and decreasing S_g. The Donaldson et al.[23,24] data show k_{ro} increasing up to an optimum S_g value and then decreasing. This is evident for values of S_o between 30 and 60% on this Berea core. The Corey et al. correlations give isoperms which show k_{ro} to increase as S_o increases at the expense of S_g. The discrepancy between the two methods is larger at low S_o values.

In the second comparison, data of Schneider and Owens[25] have been used to obtain

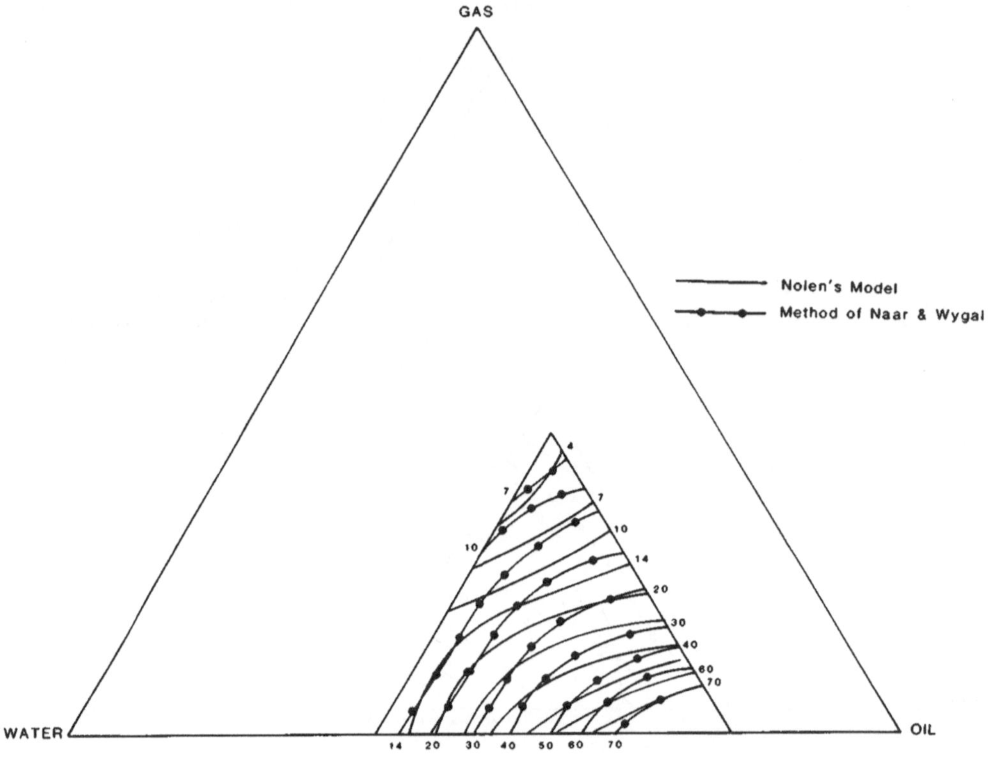

FIGURE 26. Comparison of three-phase oil relative permeability determinations.

isoperms by Nolen's model[19] and by Naar and Wygal's correlation.[12] Few data are available in the literature that show how the latter method compares with experimental values or other correlations. Figure 26, however, provides such a comparison. Schneider and Owens obtained gas-oil drainage data in the absence of connate water; their oil-water imbibition data is for a water-wet system. Theoretically, the Dietrich and Bondor[18] or the Nolen model should give the same results as Stone's second model, since gas-oil data used in this comparison have been obtained in the absence of connate water, i.e., k_{rocw} equals unity. As in the earlier comparison, the discrepancy between the two methods is evident at low S_o values. Another point to note is the evidence that k_{ro} depends only on S_o values, especially at low S_o in Naar and Wygal's correlations. There is a slight indication in both methods that k_{ro} isoperms become convex towards the 100% S_o apex at high S_g.

REFERENCES

1. **Leverett, M. S. and Lewis, W. B.,** Steady flow of gas-oil-water mixtures through unconsolidated sands, *Trans. AIME*, 142, 107, 1941.
2. **Corey, A. T., Rathjens, C. H., Henderson, J. H., and Wyllie, M. R. J.,** Three-phase relative permeability, *Trans. AIME*, 207, 349, 1956.
3. **Reid, S.,** The Flow of Three Immiscible Fluids in Porous Media, Ph.D. thesis, University of Birmingham, England 1956.
4. **Snell, R. W.,** Measurements of gas-phase saturation in a porous medium, *J. Inst. Pet.*, 45(428), 259, 1959.
5. **Snell, R. W.,** Three-phase relative permeability in an unconsolidated sand, *J. Inst. Pet.*, 48(459), 80, 1962.

6. **Snell, R. W.,** The saturation history dependence of three-phase oil relative permeability, *J. Inst. Pet.*, 59, 471, 1963.
7. **Donaldson, E. C. and Dean, G. W.,** Two- and Three-Phase Relative Permeability Studies, *U.S. Bureau of Mines*, Washington, D.C., #6826, 1966.
8. **Sarem, A. M.,** Three-phase relative permeability measurements by unsteady-state methods, *Soc. Pet. Eng. J.*, 9, 199, 1966.
9. **Saraf, D. N. and Fatt, I.,** Three-phase relative permeability measurement using a nuclear magnetic resonance technique for estimating fluid saturation, *Soc. Pet. Eng. J.*, 9, 235, 1967.
10. **Wyllie, M. R. J. and Gardner, G. H. F.,** The generalized Kozeny-Carman equation, its application to problems of multi-phase flow in porous media, *World Oil*, 146, 121, 1958.
11. **Caudle, B. H., Slobod, R. L., and Brownscombe, E. R.,** Further developments in the laboratory determination of relative permeability, *Trans. AIME*, 192, 145, 1951.
12. **Naar, J. and Wygal, R. J.,** Three-phase imbibition relative permeability, *Soc. Pet. Eng. J.*, 12, 254, 1961.
13. **Land, C. S.,** Calculation of imbibition relative permeability for two- and three-phase flow from rock properties, *Soc. Pet. Eng. J.*, 6, 149, 1968.
14. **Schneider, F. N. and Owens, W. W.,** Sandstone and carbonate two- and three-phase relative permeability characteristics, *Soc. Pet. Eng. J.*, 3, 75, 1970.
15. **Van Spronsen, E.,** Three-Phase Relative Permeability Measurements Using the Centrifuge Method, Society of Petroleum Engineers/Department of Energy, Tulsa, Okla., #10688, 1982.
16. **Stone, H. L.,** Estimation of three-phase relative permeability, *J. Pet. Tech.*, 2, 214, 1970.
17. **Stone, H. L.,** Estimation of three-phase relative permeability and residual oil data, *J. of Can. Pet. Technol.*, 12, 53, 1973.
18. **Dietrich, J. K. and Bondor, P. B.,** Three-phase oil relative permeability models, paper SPE 6044 presented at the 51st Annual Fall Technical Conference and Exhibition of the SPE, New Orleans, 1976.
19. **Molina, N. N.,** A systematic approach to the relative permeability problems in reservoir simulation, paper SPE 9234 presented at the 55th Annual Fall Technical Conference and Exhibition of the SPE, Dallas, 1980.
20. **Donaldson, E. C. and Kayser, M. B.,** Three-Phase Fluid Flow in Porous Media, DOE/BETC/IC-80/4, report published by the U.S. Department of Energy, Bartlesville, Okla., April, 1981.
21. **Manjnath, A. and Honarpour, M. M.,** Investigation of three-phase relative permeability, SPE 12915 presented at the Rocky Mountain Regional Meeting of the SPE, Casper, May 20-23, 1984.
22. **Batycky, J. P., McCaffery, F. G., Hodgous, P. K., and Fisher, D. B.,** Interpreting capillary pressures and rock wetting characteristics from unsteady-state displacement measurements, *Soc. Pet. Eng. J.*, 6, 296, 1981.
23. **Donaldson, E. C., Thomas, R. D., and Lorenz, P. B.,** Wettability determination and its effect on recovery efficiency, *Soc. Pet. Eng. J.*, 3, 13, 1969.
24. **Donaldson, E. C. and Dean, G. W.,** Two- and Three-Phase Relative Permeability Studies, U.S. Bureau of Mines, Washington, D.C., #6826, 1966.
25. **Schneider, F. N. and Owens, W. W.,** Sandstone and carbonate two- and three-phase relative permeability characteristics, *Soc. Pet. Eng. J.*, 3, 75, 1970.
26. **Fayers, F. J. and Matthews, J. D.,** Evaluation of normalized Stone's methods for estimating three-phase relative permeabilities, *Soc. Pet. Eng. J.*, 4, 224, 1984.

APPENDIX

SYMBOLS

A = area
 = constant
A_t = adhesion tension
a = material constant
B = formation volume factor
 = constant
b = material constant
C = constant
F = fraction
g = gravitational acceleration
h = thickness
I = injectivity
 = resistivity index
k = permeability
L = length
m = exponent
N = number of barrels of oil
n = exponent
P = pressure
Q = volume
q = volumetric rate
R = radius
 = resistivity
r = radius
S = saturation
 = distance in direction of flow
S^* = reduced saturation
S_L = total liquid saturation
T = time
v = velocity
Z = vertical coordinate
α = constant
β = constant
θ = angle
λ = lithology factor
μ = viscosity
σ = surface or interfacial tension
ϕ = porosity
ψ = immobile saturation

Subscripts
a = absolute
av = average
c = critical
 = capillary
cw = connate water
D = displacement
d = displacing phase
de = immobile displacing phase
e = equilibrium
 = external (radius)
 = effective
f = free
g = gas
i = initial
 = index number
 = irreducible
imb = imbibition
irr = irreducible
L = liquid
LR = residual liquid
m = minimum
mf = mud filtrate
n = nonwetting
o = oil
 = measured at 100% S_w (resistivity)
ob = trapped oil
p = produced
r = relative
 = residual
s = solution
SL = total liquid
STD = standard condition
T = total
t = trapped
w = water
 = well
wt = wetting
xo = flushed zone

INDEX

A

B

C

D

X

Y